FROM THE GROUND UP

Also by Stephanie Anderson

One Size Fits None:
A Farm Girl's Search for the Promise of Regenerative Agriculture

FROM THE GROUND UP

THE WOMEN REVOLUTIONIZING
REGENERATIVE AGRICULTURE

STEPHANIE ANDERSON

THE
NEW
PRESS

NEW YORK
LONDON

Portions of this manuscript originally appeared in slightly different form as "Disturbance"
in *Ninth Letter* 19, no. 2 (Fall/Winter 2022–23).

Requests for permission to reproduce selections from this book should be made through
our website: https://thenewpress.com/contact.

Published in the United States by The New Press, New York, 2024
Distributed by Two Rivers Distribution

ISBN 978-1-62097-814-6 (hc)
ISBN 978-1-62097-894-8 (ebook)
CIP data is available

The New Press publishes books that promote and enrich public discussion and
understanding of the issues vital to our democracy and to a more equitable world.
These books are made possible by the enthusiasm of our readers; the support of a
committed group of donors, large and small; the collaboration of our many partners in
the independent media and the not-for-profit sector; booksellers, who often hand-sell
New Press books; librarians; and above all by our authors.

www.thenewpress.com

Book design and composition by Bookbright Media
This book was set in Bembo and Oswald

Printed in the United States of America

10 9 8 7 6 5 4 3 2 1

Contents

Introduction

March 2020: the month when the world as most Americans knew it ended. A new reality took its place, one clouded by fear and uncertainty. Almost everyone fortunate enough to survive the COVID-19 pandemic has a story about those early days. Some tales are dramatic—nurses and doctors toiling in overwhelmed hospitals; elected leaders facing tough decisions. Some are filled with danger. Here I am thinking of workers forced without protection to cut meat or pack boxes or prepare food. Other stories are modest, characterized by days that swung wildly from stressful to boring and back again. Mine is that kind of story.

The university where I taught English and creative writing moved its classes online, and like thousands of other teachers I pivoted to virtual lessons. I graded papers and delivered lectures and met with students from behind a laptop—I was safe, but logging the longest days of my academic life creating online courses from scratch. My husband, who works in the investment industry, guided clients through what can only be called a stock market meltdown. We acquired masks and hand sanitizer, toilet paper and bleach. The future receded from focus; the day-to-day demanded center stage.

We second-guessed everything. Should we come back to the store when it's less busy? Should we wear two masks or one? Is my cough a late spring cold or something more sinister? Will I lose my teaching job when the semester ends, and will my husband's business survive? Inside those precarious, narrow days, we intentionally created joy. We hiked the Appalachian Mountains with our greyhound, ordered takeout, laughed through Zoom happy hours with family and friends, and watched an unhealthy amount of television. We felt thankful to be at home and alive, yet we knew sickness or death could come for us or our loved ones any day.

The intensity of our jobs and our pointed pursuit of off-hours fun were both orchestrated distractions from our fears. For it seemed we were catching a glimpse of an apocalypse: a deadly, highly communicable virus with no cure, shortages of medical personnel and equipment, an economic collapse, and mass unemployment. To many people living in safe parts of the world, in secure times of relative plenty, this foretaste was terrifying.

And to make matters scarier, grocery store shelves emptied. Ordinary goods like rice, canned goods, and flour disappeared. Occasionally people fought over what remained. I remember seeing the produce section at Ingles, a grocery store chain in western North Carolina, as picked over as the stands of a farmers' market in its final hour. In the meat section, signs limited customers to two packages of chicken and ground beef, but the cold cases were usually empty. Americans had not experienced food shortages like this since World War II; most had weathered at worst temporary scarcities, such as before and after a hurricane or blizzard. Seeing bare shelves touched a fear almost as intense as death by coronavirus: starvation.

But to say we had a "food shortage" isn't true. Americans were never at risk of starving, even when meat-processing plants shut down, farms furloughed workers, and producers declared bankruptcy. Shortages happened in part because we panic bought—and with shame I include myself in that group—but mostly because the top-heavy processing and distribution systems built around industrial agriculture failed to adapt to changing markets and shuttle food where consumers needed it. The *Titanic*-size ship that is the conventional food industry could not swerve to avoid a pandemic-size iceberg. The system could not, for instance, quickly repackage and reroute goods originally destined for restaurants and schools to grocery stores, food banks, and small community outlets instead. Meanwhile, farmers let vegetables rot in the fields, dairy producers dumped milk, and factory chicken and hog farms euthanized animals—all perishable or time-sensitive products—because the supply chain couldn't absorb them in time. Such waste was one of the most glaring signs of how exposed producers are to industry shocks. The result of the chaos: the house of cards that is America's food system fell swiftly and painfully.

Before we go any further, we must define a basic term: food system. According to the United Nations Food and Agriculture Organization, food systems "encompass the entire range of actors and their interlinked value-adding activities involved in the production, aggregation, processing, distribution, consumption and disposal of food products that originate from agriculture, forestry or fisheries, and parts of the broader economic, societal and natural environments in which they are embedded. The food system is composed of sub-systems (e.g., farming system, waste management system, input supply system, etc.) and interacts with other key systems (e.g., energy system, trade system, health system, etc.)."[1] In other words, the food system is the land and sea and air, the plate of food on the table, the landfill, and everything in between. It's the obvious (wheat in a field or cereal boxes on a shelf) as well as the obscure (the farm loan that paid for the wheat and the trees from which the cereal boxes came). The food system has too many components to name individually, but they work more or less together.

Except if they don't. What we as a nation saw in 2020, many of us for the first time, was the brittle nature of America's industrialized, consolidated food system. We witnessed the vulnerability of national distribution chains and farms too massive and specialized to change their practices. We recognized that too few big players controlled too much of the food supply. As individuals, we realized that we had handed corporations almost complete control over what we ate, and that socially and economically vulnerable people provided our food at enormous risk to their health and safety. We felt the danger of not knowing our farmers and ranchers, and of not having robust local and regional foodways in place. People may not have had the facts and figures behind what they saw, but the evidence was on display. We *felt* it.

Consider the grain sector, where almost 90 percent of the global grain trade is in the hands of Cargill, Archer-Daniels-Midland, Bunge, and Louis Dreyfus.[2] In the food service management industry—think restaurants and cafeterias in schools and universities, government buildings, stadiums, corporate headquarters, and such—Sodexo, Aramark, and Compass Group control almost 80 percent of food purchases.[3] Food processing is also highly centralized. Among fifty-five grocery categories representing products Americans typically buy, just eight categories are highly competitive.[4]

Look to your Sunday football spread for an example of the consolidation; 60 percent of chips and 80 percent of dips come from PepsiCo.[5] Just four massive companies—JBS, Tyson Foods, Cargill, and National Beef Packing Company—control 85 percent of the U.S. beef market that supplies burgers.[6] Conditions are not much better for chicken and pork, with the largest four companies in each industry carrying out 70 percent of hog processing and 54 percent of poultry processing, respectively.[7] You may also see evidence of consolidation in what you feed your children. Three companies sell 95 percent of powdered baby formula, and one company, Danone, churns out 80 percent of soy milk.[8]

The relatively small number of conglomerates that dominate the food industry source almost exclusively from large conventional farms via production contracts. Corporations have the monopolistic power to set artificially low prices, leaving farmers with little recourse other than merging with larger operations—or bankruptcy. Under these economic conditions, farmland continues to be consolidated into fewer hands each decade, which means a smaller number of ever-bigger farms grow most of what we eat.[9] In fact, the largest 4 percent of U.S. farms account for more than two-thirds of all agricultural production.[10] And then there are grocery stores, which are also consolidated. In 2019, Americans bought 69 percent of their groceries from the four largest retailers (by the way, Walmart on its own controls 25 percent of the total grocery market).[11]

In these concentrated conditions, workers have little power to secure safe conditions, fair pay, benefits, and job security in one of the most dangerous industries in America. The agriculture sector has long relied on immigrant farmworkers—vulnerable people who may not speak English or have legal residency status, permanent housing, and resources to organize. Excessive heat, repetitive movement injuries, long hours, chemical exposure, respiratory issues, skin disorders, infectious diseases, and hazardous machines are just a few dangers these workers face. Most earn little more than the minimum wage, lack insurance and transportation, live in substandard housing, and endure food insecurity.[12] Forced labor, debt bondage, and human trafficking plague the system as well, with my adopted home of Florida being a routine offender.[13]

Americans saw the brutal nature of food system work up close during the pandemic. For instance, meatpacking companies disregarded worker safety and downplayed COVID illnesses and deaths among their workforces. Tyson's legal team even drafted an executive order to keep meat plants open, a slightly reworded version of which the then president went on to sign.[14] Food system workers who can barely survive their jobs simply do not have the ability to face off against Big Food corporations that control the industry.

As the pandemic receded, a new stressor emerged: inflation. Record-high inflation hit world economies in 2022, sending U.S. food prices up 10.4 percent in December 2022 compared to December 2021.[15] Big Food claimed increased prices were necessary to cover rising commodity, energy, and labor costs. But profit margins at many food companies widened in 2021 and 2022, suggesting that price hikes more than covered costs and perhaps hinged on greed instead.[16] Shoppers remain at the mercy of this corporate food system, with the poor and socially disadvantaged bearing the heaviest financial burden of inflated prices.[17]

Of course, disruption is normal during an emergency. Thinking that any food system will remain fully intact when shocks hit is unreasonable. I don't expect perfection, and I imagine consumers and those working in the food system don't either. Resiliency and stability *are* reasonable to expect, though. Because here's the thing: the pandemic was bad for food and agriculture, but much of the "bad" was of the industry's own making. Often in the name of efficiency and profit, the food system set itself up to fail in moments of crisis, despite warnings from food and agriculture researchers, economists, even farmers and ranchers. The industry lacks diversity and true competition, and it is too centralized. Its growers, processors, and distributors are too big. It's no wonder such a system collapsed during the pandemic and took a long time to recover.

To achieve adaptive resilience to shocks, we need a more balanced combination of local, regional, national, and global supply chains, ideally powered by what is known as regenerative agriculture. Regenerative agriculture means farming and ranching practices that restore organic matter and biodiversity in soil, which in turn improves water cycles, fuels life above ground, and increases crop resilience and nutrient density. What's more, regeneratively managed soil draws down and stores more carbon compared to conventionally managed soil, and regenerative producers rely far less on fossil fuels. Both attributes make regenerative agriculture a crucial way to reduce CO_2 emissions.

Regenerative producers also take a holistic approach to land management. Rather than chasing narrow goals like yield per acre, they prioritize the ecosystem's overall health, an ecosystem that includes human communities damaged by industrial agriculture. For instance, regenerative producers typically rotate livestock across the land, a technique that builds soil health and contributes to clean air and water, whereas conventional producers may confine animals in feedlots that sicken the surrounding human and more-than-human community.

This kind of ecosystem health stems from the fact that regenerative practices are based on nature's processes. For example, no-till/ minimum tillage replicates the unbroken (and thus immensely fertile and carbon-storing) soil of the historic American grassland. As I argued in my first book, regenerative agriculture is a one-size-fits-none philosophy, not a one-size-fits-all approach like conventional agriculture often is. A cover crop that works in one farmer's rotation may not work in another's, for example, because of the farm's annual rainfall or soil profile. By favoring adaptability over conformity, regenerative operations tend to be more diverse. Diversity is a foundational principle in nature that regenerative agriculture mimics—and that diversity makes regenerative operations more resilient in changing conditions. A broader food system with comparable diversity—one that sources from a wider network of farms, builds in healthy supply chain redundancy, and demands regeneratively instead of industrially grown products—would also be more resilient.

Instead, we have an agricultural system comprised almost entirely of industrial operations. Food and agriculture companies have spent decades consolidating. Local and regional foodways have fragmented, allowing national and global food chains to dominate. Everyday farmers and ranchers are prisoners in this system; they are all but forced to mold their operations to the food industry's demands and are incentivized to do so with government aid. As the consumer advocacy group Food and Water Watch describes it, "This is not a broken system. It is functioning as it was designed: to funnel wealth from local communities into the hands of corporate shareholders and executives."[18]

The pandemic did not create the conditions for our food system's failure. It just exposed and worsened them.

I began to interpret our food system's reaction to COVID-19 as a foreshadower of how it might respond to escalating climate change–related shocks.[19] As the breakdown unfolded over months, then years, the pandemic looked more and more like an example of how extreme events, in this case a virus with cascading consequences throughout society, can topple shaky social structures.

Closed businesses, lost revenue, and unemployment seemed akin to what farmers and ranchers suffer when multiple weather emergencies strike at once, like Texas producers did when a drought, hurricane, and freak freeze occurred within the same year.[20] The pandemic's food supply and demand mismatches looked similar to how steady warming is shifting agricultural production zones and prompting other kinds of supply and demand mismatches.[21] Crops once suited to a given region fail more often now (that is, the supply changes), yet the industrial farms, businesses, and production chains built around those crops struggle to adjust course (that is, the demand does not match the supply). News about food system workers falling ill and dying sounded like reports of climate refugees enduring harrowing journeys and perishing in the Mediterranean Sea, Darién Gap, and Rio Grande. The "new normal" we resisted during the pandemic, societal transformations so big we struggled to grasp them, felt like rising sea levels, which also boggle the mind but for which we must prepare.

A correlation also exists between the ripple effects of the coronavirus and dwindling water supplies. Water levels are already dangerously low throughout the West, particularly in California and in states reliant on the Ogallala Aquifer and the Colorado River. Population growth and overdevelopment deplete water, but so does climate change via more evaporation, decreased snowpack, and floods that contaminate water. It doesn't help that farmers and cities draw more water as climate change intensifies droughts and heat waves. Agricultural areas created around irrigation find themselves rationing water or being denied it altogether. Without enough water, producers are plowing crops under, leaving fields unplanted, selling livestock, even going out of business—all of which threaten the food supply. Each time a water source goes offline, negative consequences ripple through the agricultural community, similar to how each time COVID cases spiked, problems followed in the form of school and business closures, overrun hospitals, and deaths.

Another pandemic / climate change commonality: positive feedback loops. Such loops occur when the outcome of an action or event triggers more of that action or event, and too little negative feedback tempers the effect. For example, Black families went into the pandemic with fewer financial resources on average than white families, in part because of decades-long labor market discrimination and education disparities that tend to funnel Black workers into low-wage jobs like food service, health care, and retail. When the pandemic hit, Black workers lost their jobs at higher rates than white workers as their retail workplaces closed. Workers of color also died at higher rates than white workers because of their overrepresentation in front-line jobs like food service and health care.[22] Job loss and death made possible by labor market discrimination and education inequality put affected families even further behind financially—a positive feedback loop. The federal government provided some negative feedback through stimulus checks, a pause on student loan payments, and other relief programs, but not enough to counteract all losses. And nothing can counteract the loss of life.

Feedback loops look similar in a climate context. Here is one example: The outcome of releasing greenhouse gases is higher temperatures, which melt polar ice caps. Ice usually reflects solar radiation and cools the planet—but as the ice shrinks, so does its cooling power, which causes the planet to heat even more. More heat in turn melts more ice. The cycle intensifies in the absence of negative feedback, like greenhouse gas reductions.[23]

Droughts can be another self-perpetuating cycle, like the megadrought that started in 2000 in southwestern North America. Drought makes soils drier, which means less evaporation. A lower water supply in the air means less rain, which worsens the drought, and the cycle continues.[24] And just like the pandemic changed aspects of American society forever, the Southwest's driest period since 800 CE is permanently altering the region's agriculture as farms go out of business or quit growing crops they once did. It's worth noting that human-induced climate change fueled 40 percent of the Southwest drought and, as scientists put it, "without [anthropogenic climate change], 2000–2021 would not even be classified as a single extended drought event."[25] Left unchecked, positive feedback loops can lead to tipping points, or thresholds at which change becomes self-perpetuating and essentially irreversible.[26]

Observing the positive feedback loops during the pandemic, I imagined similar loops and tipping points playing out within the food system as climate change accelerates. For instance, conventional farmers and ranchers now face more frequent and damaging natural disasters, yet their practices make them more vulnerable to those events *and* contribute to the warming behind them. Take agrochemical use, a typical conventional practice: synthetic fertilizers, herbicides, and pesticides decrease organic matter and kill soil biology, which reduce the soil's ability to absorb water during heavy rains and floods and to hold water during droughts. Instead of using regenerative practices to strengthen soil and make it more resilient during extreme weather, conventional farmers increase their soil's vulnerability with agrochemicals (and other industrial practices like monocultures and tillage, but let's focus on chemicals for now). These fossil fuel–based agrochemicals and the machines that apply them release greenhouse gases, which worsens warming, which prompts more extreme events that erode weak soil and cause crop failures. If the cycle continues, the soil will be too degraded for production—a tipping point reached on the Great Plains during the Dust Bowl, but one that would be far more difficult to recover from in today's hotter climate.

The potential for ecosystem failure is especially troubling, because failures in nature negatively impact food production. People tend to think of farming and nature as separate, but the reverse is actually true. Both industrial and regenerative food production rely on ecosystem services: water and nutrient cycling, pest and disease regulation, waste management, flood and erosion control, air quality. More than 75 percent of leading global crops rely wholly or in part on pollinator species like bees, butterflies, bats, and birds.[27]

But the planet's biodiversity is plummeting; the species extinction rate is at least tens to hundreds of times higher now than the 10-million-year average.[28] Insects have been hit especially hard, with 40 percent of species under threat of extinction.[29] That spells trouble for pollinators and, in turn, food production. Changes in land and sea use are the main cause behind biodiversity loss, with agricultural expansion being the most significant form of land use change.[30] Biodiversity loss and other seismic environmental shifts can lead to virtually permanent ecosystem collapses that make food production incredibly difficult in the affected regions.

The pandemic was more than a foreshadow, I realized. It was proof that America needs an entirely new food system, one that is greener, more resilient, and more equitable. A system that doesn't disintegrate, but regenerates.

Before the pandemic, I thought of "regenerative" mostly in terms of its agricultural practices for growing crops and livestock. And when it came to the people involved in the food system, I mostly studied farmers and ranchers. A sort of legitimacy scale dominated my thinking, with people on the land at the top and everyone else—policymakers, advocates, researchers, agriculture-related businesses, investors, consumers, myself as an agricultural writer—below them.

Maybe my background led me to see producers and their work as the most important parts of regenerative food production. I'm a daughter of farmers and ranchers; I worked cattle, drove grain trucks, and disked fields. I spent far more hours on horseback than

I did in front of a television. My mom, who could rope and ride with the best of them, taught me and my siblings how to tighten a latigo just right, how to stay on a green horse that bucked. When I wasn't on a horse, I was fixing barbed-wire fences, bottle-feeding calves, shoveling shit out of trailers, or crawling across the prairie on a swather. Summer meant waking up before the sun to rake hay, fluffing up the windrows for my father to bale, the damp alfalfa and crested wheatgrass and sage perfuming the air. I am immensely proud of my parents, who grew our ranch and farm into a prosperous operation, even if I disagree with some of the decisions behind that success and feel shame about the settler colonialism that put stolen Indigenous land into my ancestor's hands. Still, when I think about food and farming, I picture my parents and people like them.

So when I write about regenerative agriculture, I tend to approach the subject from a producer's perspective. Sure, I am concerned about the broader food system, but researching and writing about it meaningfully felt out of my wheelhouse for years and, I admit, less exciting than stories about ranchers and farmers. I figured someone more familiar with farm policy and agriculture economics and nutritional science deserved to write about those subjects. But as was its wont, the pandemic changed everything.

I came to the understanding (rather obvious, I think now) that regenerative agricultural practices are a beginning, not an end, in constructing a resilient food system. Compared to conventional production, regenerative practices lead to so many benefits: carbon sequestration, lower greenhouse gas emissions, healthier soil and ecosystems, cleaner water, increased biodiversity, more robust rural communities, and more nutritious food. Regenerative operations are better positioned to handle the consequences of climate change and maintain a stable and affordable food supply. They tend to be more profitable; regenerative corn production, for instance, is

78 percent more profitable than conventional, and that is without much of the government and institutional aid that conventional producers enjoy.[31] Those outcomes would undoubtedly make our food system stronger overall and more flexible in times of change. But consumers will always be at risk if the rest of the food system remains consolidated, focused on cheapness over quality, and rooted in national and global rather than regional and local supply chains—all propped up by agribusiness and government.

I knew regenerative agriculture was gaining ground among farmers and their advocates. Yet the food system as a whole seemed intractable, before the pandemic at least, and I had little sense of whether change was possible. Were the images on TV and the empty shelves around me part of a predetermined future for our food system, or could we change those scenes that promised to unfold again when the next disaster struck? Once travel was safe, I embarked on a journey to find out—and I discovered that we do indeed have a blueprint for a better system, created by a rising collective of diverse female farmers, entrepreneurs, community organizers, scientists, investors, and policy advocates.

Women-run farms and ranches across America are upending conventional agriculture and going regenerative instead, with support from women-led investment funds, training programs, restaurants, food brands, and advocacy groups. Women are finding ways to connect farmers and consumers through direct marketing, such as farmers' markets, food hubs, and community supported agriculture (CSA) programs, and making our food system more local and regional as a result. Women entrepreneurs are creating regenerative food labels and products and establishing distribution chains linking small and midsize farmers to buyers. Women in scientific and political leadership positions are advancing regenerative agriculture as a way to help address environmental, social, and human health issues. All these combined efforts are disrupting the conventional status quo and spurring progress within the food system.

This work began well before the pandemic. Many female farmers and ranchers have long understood that our food system is broken. So did many women running food companies, farmers' markets, investment firms, grocery stores, research labs, government programs, and agricultural support organizations. An urgent approaching trouble was, and still is, climate change, which promises disturbances far more intense than a global pandemic. My journey into the depths of the food system indicates that a growing number of women are answering the call to create a regenerative food chain in time to avoid the worst effects of climate change and other disruptions the world might throw our way—and that is what this book is about.

Women are joining the regenerative movement in response not just to climate change, but also to social inequity, declining human health (especially in children), environmental degradation, and rural economic stagnation. Because women often view the world through the context of relationships, many are gifted at thinking holistically. They understand that regenerative agriculture can restore human and biological health together, and revitalize communities and small businesses. It can help address social injustices and promote diversity, equity, and inclusion. And it creates opportunities for the next generation. Research shows that women tend to approach problem-solving through collaboration; they lead with empathy rather than domination. Regenerative agriculture is the ultimate form of collaboration and empathy between people and nature, a partnership many women in the regenerative movement see as vital to the survival of both.

While many men share this green vision, and are typically credited for "inventing" regenerative agriculture, the sheer number of women involved in redesigning our food system is altering a histor-ically male-dominated profession. Their ecological consciousness sets them apart from colleagues who control and lead the conventional, environmentally extractive food chain. This wave of female leaders is diverse in race and ethnicity, age, crop and business type, geographic region, sexual orientation, and personal background, which challenges the white male farmer and businessman stereotypes. Many of these women expand the definition of regenerative to include equity and diversity, social and environmental justice, and success for young farmers. If conventional food is the still-water line, the status quo, then these women are the energy moving the water.

This book focuses on women not only because of their contributions to the regenerative movement, but also because conveying their experiences will, I hope, help shift notions about who belongs in the food system. Mythologies about the white male farmer remain incredibly persistent. Such thinking led, in part, to the extractive, industrial agricultural world we have today—a world built by and modeled after men. Men still maintain control over many of the major companies, research projects, government agencies, and other entities that make up the food system. Women who work in that conventional system, especially nonbinary women and women of color, often face suspicion. The regenerative movement is different; it tends to be more diverse and welcoming. That may not remain the case, however, if people like me repeat the same kind of stories told about conventional agriculture.

But women cannot save the food system alone. We must apply their thinking across the entire food chain, and quickly, to avoid the worst effects of climate change. Food shortages, crop failures, livestock losses, ecosystem collapses, skyrocketing grocery store prices, and hunger likely await us if we fail. This book will show you some of the people doing and applying that thinking. The work is theirs, not mine; I am simply telling their stories. In sharing these accounts, I do not intend to perpetuate a hero narrative in which a select group of individuals deserve all the credit for building a sustainable food system. That kind of savior story does not reflect the collaborative nature of the regenerative food movement, or the growing number of people within it, or even the women featured here, whom I suspect would reject the notion that they are heroes. Rather, my goal is to share examples and inspire change and participation through narrative. The women here and beyond are pioneering solutions, providing vision, and proving that regenerative eating is possible—and if we follow their lead on a broader scale, then our food system will be more prepared for a changing climate.

1

Disturbance

Kelsey Scott, DX Beef

Disturbance: the agitation or disruption of a steady state.

I am privileged that fear isn't an emotion I associate with my childhood home in western South Dakota. But in late September 2021, for the first time, fear takes my hand from the moment I leave the Pierre airport and drive west on Highway 34. I see hollow depressions where water should stand and bare ground where grass and forbs should grow. I see herds of cattle packed together, fighting flies and kicking up dust. Some pastures are so short they look as if ranchers cut them for hay. Each scene seems a foreshadower to the future our scientists say is coming, a future that in many ways already exists. During the last years I lived on my parents' ranch, in high school, drought ruled. I am no stranger to loss and lungs filled with dirt. Neither is the prairie itself, forged from cycles of wet and dry, plenty and little, activity and rest—a resilient land that gave life to Indigenous tribes and countless species of animals, plants, and microfauna for millennia. But this is not the prairie ecosystem of old, and this is not the climate it once knew.

At sunset I cross into tribal lands at the Cheyenne River valley and think about my plans to return to this reservation in a few days to interview a young woman rancher of the Lakota Nation. For a while these thoughts carry me up Highway 73 in peace. But when I am less than an hour from home and the world is dark, my rental car's headlights illuminate a dead deer on the shoulder—a deer that turns out not to be dead after all, but injured from a car collision, and which heaves itself halfway up like a zombie right as I pass. At the end of the three-hour drive my teeth are clenched, shoulders stiff, eyes dry from watching for movement in the ditches.

The tension carries into the days with my family. It's not a tension between us—my mother, father, brother, and I share meals, catch up after months apart, laugh together—but a tension emanating from land in distress. It's also tension around loss. Ranchers across the Great Plains are selling cattle, running out of pasture forage, and harvesting drastically less hay than usual, if any at all.[1] My dad, a conventional rancher and farmer, put up a quarter of the hay he typically harvests over a summer. The wheat, oat, and corn crops are not great either. One afternoon, my mother, brother, and I drive to a pasture to make sure the cattle's water tanks are working. The herd mopes around the water; a dead calf bloats nearby, victim of dust pneumonia. The cows are skinnier than I have ever seen in our herd. The ash trees in the surrounding draws glow orange and yellow, absurdly beautiful in the otherwise bleak scene. "This is a disaster," I say, less a comment to my mother and brother and more an apology to the land.

That day, I long for the prairie that once was, a prairie I have never seen though I lived on a High Plains ranch for eighteen years. North America once had three types of prairie: tallgrass, mixed grass, and short-grass.[2] The tallgrass prairie spanned the eastern edges of what is now the Dakotas, Nebraska, and Kansas; most of Iowa and Illinois; and swaths of southwestern Minnesota, northern Missouri, central Oklahoma, and eastern Texas. Moving west, tallgrass fringed into mixed grass (a blend of tall- and short-grass species), which in turn gave way to short-grass that stretched from the western halves of the Dakotas, Nebraska, Kansas, Oklahoma, and Texas all the way to the Rocky Mountains. The three prairies supported a complex array of animals, plants, insects, and humans working together. Herbivores like elk, pronghorn antelope, and white-tailed deer grazed the land. These herbivores encouraged plant diversity, tramped down dead growth, made space for species that prefer open ground (like the burrowing owl), fertilized and aerated soil, and supported predators (like wolves and mountain lions).[3] Prairie dogs constructed burrows, which European colonizers called towns, that sometimes collectively stretched hundreds or thousands of square miles.[4] Prairie dogs, too, foraged, mineralized, created habitat, and served as food sources for other species. Herbivores flocked to the protein-rich fodder that grows in prairie dog towns, as livestock do today.[5]

Bison were the prairie's main animal architects, however. In massive herds, they disturbed areas aggressively and then migrated on, usually not returning for a long time given their vast territory. They ate from, wallowed upon, crushed down, excreted on, and churned up the land. Vegetation grew back stronger after the disturbance, and the bison's extended absence allowed habitat regrowth for species like the greater prairie chicken that require cover. Fire played a similar role, and it was especially crucial in controlling woody plants like eastern red cedar, dogwood, and sumac on tallgrass prairie, which almost always reverts to forest without fire disturbance.[6] The dense, nutritious plant growth after fires attracted grazers, grazing being another way nature kept woody plants in check. Knowing this, Indigenous tribes set fires to lure animals for hunting and regeneratively manage plant communities as needed.[7] Natural and human-ignited fires reduced the prairie's flammable material, known as the "fire load," which prevented blazes from becoming overly large and hot—as opposed to the wildfires we see today after decades of fire suppression.

I witnessed the significance of grazing and fire at Konza Prairie Biological Station in the Flint Hills of Kansas. In summer 2021, I took a tour with John Blair, the station's director and a co-investigator on its Long-Term Ecological Research program. Konza Prairie is a more than 8,600-acre tract of native tallgrass prairie, named after the Indigenous Kaw people, who stewarded the land until the U.S. government forcibly removed them in the 1800s. Since 1971, Konza Prairie has served as a research site, notably for long-term studies that are increasingly vital for understanding how climate change impacts land over time.

John and his team analyze how prairie reacts to bison or cattle grazing, fire regimes, combinations of the two, and neither, observing how climatic conditions interplay with those forces over the years as well. He shows me two tracts of land that had similar prairie vegetation profiles when a fire versus no fire experiment began in 1971. The tract subjected to periodic burning contains diverse plant species and very little woody vegetation; it looks like a prairie. The unburned land, however, has morphed into a cedar forest so impenetrable that researchers can barely walk through it. "In an area like this, where the climate is sufficient for multiple vegetation types, you can think about there being alternate stable states. One of them is the prairie," Blair explains. "Prairie requires a certain level of disturbance. It requires relatively frequent fire, periodic droughts, the presence of grazers, to be a functioning prairie. And if you remove those things, then you're likely to move it into one of these other alternate states, which, depending on how deep the soils are and how wet it is, might be a shrubland, or it might be a cedar forest."

So disturbance is essential, as long as it occurs neither too often nor too infrequently. Another important benefit of disturbance occurs underground, down in the soil. Aboveground pressure from grazers and fires forces plants to create immense root systems to store energy, in the form of carbon, for recovery.[8] That means prairies sequester carbon, a function our warming planet desperately needs. Prairies recovering from poor management can be "carbon sinks," or landscapes that store more carbon than they emit, as their carbon returns to pre-colonization levels. Once prairie soils reach their carbon threshold, they become carbon neutral. Even appropriately burned prairie is good from a greenhouse gas perspective, as Blair explained.

"When you burn the prairie, the aboveground biomass is being converted back to carbon dioxide, but burning increases the growth the next year, both above- and belowground. It increases the growth of roots as well as the growth of the shoots of the plant," Blair says. "And so that offsets, more than offsets, the loss of CO_2 in fire. Burning is a positive aspect for maintaining prairies." Those carbon-storing roots also hold the soil down to prevent wind and water erosion and feed a lively biological community underground that enriches plants. That organic matter also helps the land absorb water during extreme weather events and hang on to it during droughts. Each 1 percent increase in soil organic matter equals twenty thousand more gallons of water storage per acre on average—resiliency embedded from the ground up.[9]

Tallgrass prairie is functionally extinct, plowed up for agriculture and converted to cities. Just 4 percent of the original 170 million acres remain, and those are located mostly in small, scattered, and unprotected tracts, except for preserves in the Flint and Osage Hills of Kansas and Oklahoma.[10] Konza Prairie is one of those preserves. About 30 percent of our mixed-grass prairie[11] and 70 percent of our short-grass prairie[12] are left, but in a degraded state, mainly because the prairies' original human and animal caretakers were pushed out. In a systemic genocidal campaign, the U.S. government massacred the Indigenous people who tended the land and forced survivors onto reservations, limiting to this day the acres they can steward.[13] With government support, white colonizers and hunters exterminated the bison, wolves, and elk; almost eliminated many critical species such as black-footed ferrets and mountain lions; and drove down countless other wildlife populations. They dammed rivers and creeks, restricted animal movement with fences, and brought land management attitudes rooted in dominion, racism, and sexism.[14] And they practiced conventional grazing, now the most common form of livestock management in the western United States.

Conventional grazing typically means keeping livestock in large pastures for long durations. Unless encouraged to move, domestic livestock tend to repeatedly graze the same easy-to-reach areas, often near water or protection from the elements. This clustering causes uneven fertilization. It also leads to overgrazing in choice areas that weakens and eventually kills perennial plants and releases carbon, leaving bare ground behind. Bare ground becomes so hot that soil biology dies and organic matter declines. Too little grazing in other places allows dead plant matter to accumulate and chokes out new growth, also killing plants.[15] Such grazing deteriorates plant communities by reducing diversity and allowing introduced species like crested wheatgrass and smooth bromegrass to take over. The most irresponsible conventional grazers keep livestock in the same pasture for so long that the whole tract becomes overgrazed.

Conventionally grazed cattle are the fuel that powers America's Big Beef monopoly. Just four massive companies—JBS, Tyson Foods, Cargill, and National Beef Packing Company—control

85 percent of U.S. beef packing, a term that covers slaughter, cutting, packaging, and distribution.[16] Given the consolidation, ranchers have virtually no choice but to sell their livestock into this system. Even organic, grass-fed animals often wind up in Big Beef feedlots before slaughter, since those companies now control a number of organic and grass-fed beef labels. The Big Four use their monopoly to drive down cattle prices, which in turn drives ranchers out of business. As one analysis notes, "Today, in many regions of the country, ranchers report finding as few as two buyers in a market, and increasingly these buyers do not compete against one another. One result of this consolidation is that the price ranchers receive continues to fall, even though the price consumers pay for beef is rising and beef packers are making record margins."[17]

Now the Big Four's sights are set on the feedlot industry. Independent feedlots traditionally serve as a middle step between ranchers and beef-packing companies. Feedlots source cattle from ranchers, usually via live auctions at sale barns but also through private contracts. These individual, often family-owned feedlots fatten the cattle in confinement for several months, then sell them to packers for slaughter and processing. But over the years the packers have vertically integrated by acquiring their own feedlots and pushing independent operators out. In the past twenty-five years, 85,000 independent feedlots left the industry—that is 75 percent of the nation's cattle feeders gone, most of them families. The packers then ballooned many of their feedlots into massive concentrated animal feeding operations across the West. Today the largest seventy-seven feedlots house 35 percent of all fed cattle in the United States.[18] The few independent feedlots that are left have dwindling options: they can sell slaughter-ready cattle to the Big Four, attempt to contract with a smaller packer (which can be difficult to do)—or go out of business.

To be clear, I am not advocating a return of independent feed-lots, family or corporate. Feedlots and other confinements are completely unnecessary for bringing livestock to slaughter weight, as any producer of grass-fed animals will confirm. Proponents argue that animals reach slaughter weight faster with less human labor in feedlots, efficiencies that seem to justify the system. But animals only put on weight so fast because they are fed outsize quantities of corn- and soy-heavy rations, the equivalent of junk food. Gaining weight quickly is hard on animals, as with the chickens whose breasts get so heavy in confinement that they can't stand. Feedlot rations are especially bad for cattle, ruminant animals evolved to eat grass that become sick on constant grain. Confinements prevent animals from moving, too, which speeds up weight gain but at the cost of a humane life. Consider the gestation crates that pregnant sows in many U.S. states live in, crates so small they cannot even turn around.

Feedlots and similar livestock confinements have many human costs as well. They cause disastrous air and water pollution that sickens people, serve as vectors for disease-carrying insects and pathogens, and emit intolerable odors. Many feedlot animals receive continuous low-level antibiotics, and we know that overuse of antibiotics leads to antibiotic-resistant bacteria.[19] Feedlot-raised meat and milk are also less nutritious than grass-fed equivalents.[20] Livestock can easily achieve slaughter weight on grass and other natural rations. It may take a little longer, but giving animals more time to grow in humane conditions more than justifies getting rid of confinements.

What I am advocating is a breakup of the Big Beef monopoly. Such a move would increase industry competition and lead to fairer prices for ranchers. Whether cattle buyers represent a Big Four or independent feedlot, their hands are tied when it comes to what they can pay ranchers, thanks to the Big Four's influence. Not wanting to alienate their few remaining buyers, ranchers often feel pressured to adapt their practices to satisfy the beef industry. They might mimic feedlots by using growth hormones, low-level antibiotics, or expensive supplemental feed to increase their calves' weight. They might operate their own small-scale feedlots to make their calves more appealing to big feedlot buyers. Submitting to this system also means relinquishing control over their product. Slaughterhouses combine meat grown in this country and meat imported into this country and circulate the mixture nationwide, which means ranchers cannot trace where their products end up or who buys them. Most ranchers graze conventionally, which they believe helps their bottom line, which is ever shrinking thanks to the Big Four. The whole system depends on a prairie in crisis.

What I see out the pickup window, and what is considered "prairie" across much of the West, is a grassland ghost. A warped ecosystem broken not just by conventional grazing and farming, but also by a changing climate, discrimination, patriarchy, and a food system that demands artificial cheapness achieved through industrial practices carried out by exploited ranchers. What I see is a society-wide failure, a resilient carbon-neutral landscape made complicit in global warming. Imagine my excitement, then, when two days later I find myself walking through pastures only two and a half hours away that are thick with forage and wildlife, managed by a female descendant of the prairie's Native stewards.

This is a drought, I remind myself, running my fingers over green shoots poking through what is left of the summer's trampled grass. I am on my hands and knees next to Kelsey Scott, fourth-generation rancher and descendant of the Lakota Nation, who explains the plants before us: their medicinal and ceremonial uses in Lakota culture, their value as livestock forage, their increasing health thanks to regenerative grazing. "This is plantain," she says, touching a cluster of green leaves (these are not the banana relative, by the way). Plantain leaves are edible in many preparations but are especially useful for balms or salves, cough suppression, and digestion. "I believe I would have been one of the keepers of the plant nation if time was still of the thípi ages," Kelsey says. "I really feel I would have been one of the observational scientists that existed across my tribal peoples, as we learned how to navigate the landscape and treat from the land and heal from the land and care for the land."

Kelsey and I are at a pasture's edge on the banks of Lake Oahe in present-day South Dakota. The pasture where we sit is situated within the seven thousand acres managed as the DX Ranch, a cattle operation located on the tribal lands of the Cheyenne River Sioux Indian Reservation and run by Kelsey in partnership with her family. On this early afternoon, she wears a straw cowboy hat, geometric-patterned blue button-down shirt, charcoal work pants designed for women with a cell-phone pocket on the thigh, and worn, dust-covered cowboy boots. Her pecan-colored hair hangs in a low side braid. The lake glitters with sunlight, its azure hue almost matching the cloudless sky. Driving through the pasture to this spot, Kelsey pointed out a cluster of trees where tribal members perform the Sun Dance ceremony and a scaffold burial site on a hilltop. I feel humbled to be on this sacred land, aware of how white people figure into its history, cognizant of their brutal impact on Native people.[21]

Kelsey brought me to this lakeside pasture so I can see the results of regenerative rotational grazing. Before she and her family implemented this technique, their cattle preferred the opposite side of this paddock, where the terrain is gentler—and the herd tended to stay there. So the family divided the pasture into smaller sections, helping the cattle move across the entire area. That also encouraged the animals to graze evenly on all forage (like buckbrush, a woody plant livestock can eat but avoid in favor of grass) rather than only what they enjoy most (like little bluestem, a nutritious native grass). The family uses a combination of permanent and semipermanent electric fences to split pastures and move the herd. The cattle evenly tramp down old growth, making way for the new grass Kelsey and I observe. "I'm so excited to see what the new growth looks like this next spring, just because there's finally clearance for new growth to come in," Kelsey says, surveying the prairie around us. Regenerative ranchers call the whole process "disturbance," or concentrated short-term grazing, similar to what the bison once did on North America's grasslands, followed by long rest periods.

Kelsey, twenty-eight when I visit, is the mastermind behind many of the regenerative methods at the DX Ranch. She returned home after graduating in 2015 with a bachelor of science degree in range science from South Dakota State University, where she was the first Native American student to deliver a commencement address. She also holds a master of agriculture degree in integrated resource management from Colorado State University. Kelsey most recently continued her education as a member of the first Consciously Regenerating Ecosystems in Agriculture through Transformative Experiences (CREATE) coaching cohort. CREATE is an intensive regenerative agriculture coaching course offered by Integrity Soils, an agroecological education company run by soil health superstar Nicole Masters. In addition to ranching, Kelsey works as chief strategy officer for the Intertribal Agriculture Council, whose mission is to revitalize Indian food systems by promoting the Indian use of Indian resources.

Kelsey's holistic thinking stems from not only her education, but also her Lakota heritage. What people call regenerative agriculture is really a collection of Indigenous and Afro-American practices and philosophies, a fact too often ignored—even by me in my first book on the subject.[22] Native communities across North America, for instance, deployed regenerative methods like agroforestry, permaculture, and intercropping to grow crops and livestock.[23] They built terraces, planted riparian buffers, cultivated both domestic and wild crops, and used ash and bonemeal as natural fertilizers.[24] Aztec farmers invented the milpa, a system in which farmers grow corn along with various species of legumes, as well as squash or tubers. The corn acts as a scaffold for the beans, the beans fix nitrogen for plant growth, and the squash or tubers stretch over the soil to suppress weeds and store moisture. Known as the Three Sisters today, the practice spread to Indigenous societies throughout the Americas.[25] Mayan farmers, too, used the milpa to mimic the forest, layering root crops, the Three Sisters, small livestock, and fruit trees underneath canopies of native trees.[26] Asian farmers developed living mulch systems in rice paddies, rotated crops, composted, and integrated animals like fish and ducks into rice fields to create closed-loop farming systems.[27] Chapter 2 describes how African societies employed numerous regenerative practices like these as well.

Indigenous agriculture is a whole-ecosystem approach that balances the needs of all life, Kelsey explains, as humans work in conjunction with nature rather than against or separate from it. "Humans are a part of this system and they always have been," she said earlier this morning during a guest lecture to a group of law students that she delivered from her kitchen table as I listened in nearby. "We've been seed keepers, breeders, and pollinators as we've stewarded this land, as we've passed on Indigenous land management techniques and practices. We've been carbon and nutrient managers. We were herdsmen; we followed bison across the landscape. We managed fuel load, meaning we started intentional prescribed burns. There are Indigenous practices for brush management, for taking care in controlling burns that are now being re-explored as fire ecology in present-day natural resource management. So these are practices that have existed on this landscape. This is innate knowledge that many of our native producers carry with them."

Indigenous principles of working within the ecosystem inform Kelsey's management, with a modern spin. Put simply, she creates a series of life-giving disturbances using the tools of today.[28] Bison-inspired rotational livestock grazing is only the beginning. The family uses no broad-spectrum pesticides or herbicides on the land, and no antibiotics, hormones, or insecticides on the livestock, all of which saves money and time and protects the environment. "We want to promote healthy dung beetle activity, and expect our cattle to survive with a healthy parasite load in this ecosystem, and not just mask inefficient animals," Kelsey explains. If an animal becomes ill and cannot recover naturally, then the family treats it in a "sacrifice area," a separate small pasture, so the medicine will not interfere with insects or soil biology on the rest of the ranch. DX Ranch cattle give birth in May rather than in blizzard-prone March or April as most conventional cattle do, which also helps the ranch avoid livestock loss.

Additionally, the family strategically rotates mineral and salt across pastures to draw cattle to otherwise ignored areas, and they installed extra water tanks to enable more intentional grazing. They also practice sustainable hay harvesting by moving cattle across hayed ground in the coldest winter months when feeding bales is necessary (herds graze on pasture most of the winter, though). The cattle replace soil nutrients lost to haying by fertilizing the ground and leaving some trodden hay behind, which acts as mulch and eventually decomposes into the soil. Kelsey even "plants" native grass by mixing livestock-safe, non-chemically treated seeds into the cattle's loose mineral. The animals excrete convenient fertilizer packets (in other words, cow patties) containing seeds that germinate and grow when conditions are right.

The fresh plant growth Kelsey and I observe is normal for fall here in South Dakota—temperatures typically dip and rains revive the prairie—but this year the renewal *is* unusual. Neither the lower temperatures nor the rains came. The drought in some areas of the Dakotas is worse than the driest years of the Dust Bowl.[29] The DX Ranch caught a few fall showers, but nothing drought easing. When I check the drought monitor online, I see that this year the reservation has faced severe or extreme drought since March.[30] Yet prairie abundant with grass and forbs is everywhere on the DX Ranch, so unlike the land I observed driving between Pierre and my family's ranch. Some pastures Kelsey shows me were already grazed, but they appear untouched to my eye. "Realistically speaking, our past season's management has prepared us for this," Kelsey says when I ask about the drought's impact. "We're not going to have as much gain on the cattle naturally, we're not going to have as much standing forage left, all those things. But if we needed to harvest it, we could have just as much hay this year as we did last year, thanks to the stockpile forage we reserved from years past. It's unique because we don't feel the need to make management decisions for animals or forage we sell this fall specifically based on the drought or if the market is hot or not."

Those ranchers across the West selling off their herds—many could have avoided that outcome with regenerative management. I feel empathy for those families. Destocking is a financial and psychological gut punch; losing the ranch is far worse. Kelsey describes these losses as generational traumas that cause waves of hurt within ranching communities. Giving up the land often means giving up a hard-won family legacy, one's sense of self and purpose. The people on whose watch the ranch slips away may never forgive themselves, even if the forfeiture wasn't entirely their fault. Most farmers and ranchers understand their businesses exist in the context of uncontrollable forces: weather, global markets, fickle consumer preferences. But today's producers operate in a changing climate that promises more severe and more frequent unusual events. And the hurt those events bring about can lead to tragedy. We know, for example, that extreme weather and its devastating financial effects are one reason the farmer suicide rate in America is soaring.[31]

Transitioning to regenerative agriculture can help operations endure weather-related shocks, which could prevent those devastating waves of hurt. Regenerative practices build ecosystem resiliency so the land can better withstand the blows and recover. Regenerative operations also tend to have more crop and animal diversity, making it more likely that ranchers will stay in business if one enterprise fails. Our changing climate calls for regenerative financing and institutional support as well. Right now, most lending institutions will not fund regenerative production, or refuse to adapt lending products to better suit regenerative agriculture and the people who tend to practice it. The same is true with federal and state agriculture programs, which are too slowly reorienting to serve regenerative producers, with most aid still going to industrial production. All of this needs to change for producers to survive in a changing world.

Regenerative grazing is not a silver bullet. Some ranchers will still fail, just like people do in other occupations, even when they do everything right. But we know that conventional production is weakening the land and its producers and that regenerative grazing can reverse this damage—all while sequestering carbon to help slow the warming that causes weather extremes in the first place.[32]

Leaving the prairie empty with no human or animal involvement also is not an option. America's prairies need short-duration, high-intensity impact from herbivores to thrive and, as Kelsey and other Native scholars point out, stewardship that began with people from Indigenous tribes. That is one reason switching to lab-grown meat or eliminating meat consumption altogether are not answers to saving the prairie from climate change. Just as overgrazing weakens the land, so does neglect. The prairie needs animals and the people who manage them to withstand hotter conditions.

Kelsey and I return to her pickup to tour more pastures. Just as we climb in—and I do mean climb; it's a huge Ford-150 crew cab truck—she points to the sky over the opposite ridge. "That's interesting," she says. "It's a golden eagle pair." I spot their nut-brown bodies tilting against the aqua backdrop. "I've never seen them down on this point necessarily. So I have to imagine that the grazing clipped off enough so that they can hunt for the mice and rodents." Another benefit of regenerative grazing: symbiotic ecological relationships between wildlife, land, and livestock that mimic the bison/prairie relationship, visible today in real time.

Paying attention to wildlife patterns is another part of Kelsey's regenerative philosophy, helping her know whether and how her actions are in balance with the land, just as her Lakota ancestors did. She gathers or makes note of bird feathers, for instance, to understand what species call the pastures home. "I kind of keep track of them from a year-to-year basis to be like, how often did we have eagles around shedding feathers? It's like a management deal. The more consistently they're out there hunting, the more consistently they're going to leave evidence of it," she says. Sometimes nature's balance goes awry, though. When it comes to undesirable plants or wildlife, Kelsey again turns to ecosystem thinking to determine why Mother Nature filled a void with an unwanted species and how she might adjust her management, instead of reaching for a chemical solution. "We try really hard to wrap our mind around the idea that there's no such thing as a pest, that every animal, every species, has a role, and we just need to be better at training ourselves to interpret what that role is," she says.

Reciprocity. Respect.[33] Disturbance that regenerates. These guiding philosophies rooted in Indigenous ecology give the land before me the resilience of historical prairie. But that strength is rooted in more than management alone. There is another Native principle at work here: the valued role of women in food production.

Agriculture is not unique in its historic maleness or whiteness. Most sectors of the American economy grew up without women or people of color, and now welcome them—or try to. Given its deep patriarchal roots and notorious resistance to change, though, agriculture is one of the last frontiers of gender and racial equity. But the tide is turning. Kelsey is part of a growing movement of women into regenerative agriculture, which provides a life-promoting disturbance within the food and agriculture industry.

According to the 2017 Census of Agriculture, beginning farmers—people who've farmed ten years or fewer—accounted for 27 percent of the country's 3.4 million producers.[34] Of those beginning producers, 41 percent are women.[35] That's higher than the percentage of women farmers overall, which stands at 36 percent—a figure itself up from 30 percent in 2012, or a 26.6 percent increase.[36] In comparison, the number of male producers dropped by 1.7 percent from 2012 to 2017. Women have the highest total representation, 44 percent, among American Indian/Alaska Native producers.[37] Overall, women farm or co-farm

43 percent of the nation's farmland—that's almost 388 million acres.[38]

A new census format accounts for some of the increase in female operators—respondents could list more people involved with farm decisions in 2017 than in 2012, which meant more women (and producers generally) were counted—but leaders in food and farming say the modified question only partially explains the rise. I posed the question of whether more women were entering agriculture to Jessica Martell, former chair of the board of directors of Blue Ridge Women in Agriculture, assistant professor at Appalachian State University in Boone, North Carolina, and author of *Farm to Form: Modernist Literature and Ecologies of Food in the British Empire*. "Ultimately I would confirm, and I think farmers up here would confirm, that those trends are definitely true on the ground here," Martell says. She has seen even more noticeable increases in North Carolina's Triangle region of Raleigh, Durham, and Chapel Hill.

I put the same question to Lisa Kivirist, director of the Soil Sisters project with Renewing the Countryside and author of *Soil Sisters: A Toolkit for Women Farmers*. "The movement is much louder and broader than the data indicates," Kivirist says. "Climate change has really motivated women, I think, in ways that in reality are only going to increase, because none of this is going away. I see that, here in the Midwest for sure, increasingly new women farmers, whatever age, whatever background they may be, are going into it with a really strong stewardship ethic and ethos of caring for the land and wanting to leave it in a better situation. They're asking, what does that [land] need? And how do we do that?"

Gabrielle Roesch-McNally, Women for the Land director at American Farmland Trust, agrees that women are entering the agricultural sphere in response to environmental and climate concerns. "We can safely say that women are really passionate about these things, but I'm always careful the way I frame that because I don't want people to feel like, 'Well, it's women's job to do this work, to dig us out of this mess that we're in on the landscape.' It's everybody's, and I think there are a lot of men who are excited about it, too, but no doubt women are leading," Roesch-McNally says. Kathryn Brasier, professor of rural sociology at Pennsylvania State University, sees the rise as a confluence of better data gathering, society-wide female empowerment, and a heightened interest in food and conservation. "One reason [for the increase] is, in general, women being more active in the workforce writ large. They're increasingly doing work that historically has been seen as men's work and increasingly claiming those titles," Brasier says. "Even if they're a partner on the farm, the act of saying 'I'm a partner' versus 'I'm a farmer' is a big step for a lot of women. So I think it's both material and symbolic in the ways that those numbers are increasing."

Among farmers and ranchers practicing regenerative agriculture, women enjoy even higher representation compared to conventional agriculture—and they are infusing that regenerative sector with new ways of thinking. "Women are leading in this space of regenerative agriculture, and not just from a land perspective but in thinking holistically, in that they're thinking about community resilience and farmland resilience. They see more holistic opportunities through diversified systems, biodynamics, permaculture, and organics," Roesch-McNally explains. More women in regenerative agriculture does not mean men are not doing regenerative work. Rather, a diverse female presence distinguishes regenerative from the industrial system, which is largely controlled by white men and has been for generations. In contrast, the new wave of female producers is more diverse in age, ethnicity, crop and business type, geographic region, sexual orientation, and personal background.

"The next generation of farms, if you look at the demographics, those tend to be younger farmers, women farmers, and women who identify as BIPOC, who are leading that edge for the next generation," Roesch-McNally adds.

That's another way Kelsey embodies America's changing farmer: she's a woman of color. Of the nation's beginning farmers, the latest USDA Census of Agriculture found that just over 9 percent identify as Hispanic, American Indian, Asian, Black, Native Hawaiian, or more than one race, compared to 7.5 percent combined representation for the same groups among total U.S. farmers.[39] Admittedly, the race and ethnicity difference between beginning and total producers (9 percent versus 7.5) is not as pronounced as the gender difference (41 percent versus 36). But as Brasier points out, the Census of Agriculture tends to bypass small-scale, urban, and BIPOC producers—characteristics that are more likely among young and beginning producers.

BIPOC farmers may not respond to the census because many do not trust the USDA after decades of thoroughly documented discrimination, or they may not receive a census query at all because the USDA has struggled to reach minority farmers.[40] The Biden administration hopes to remedy both issues, but the fact remains that the latest census likely does not reflect the true diversity of new and beginning farmers. Data from the National Young Farmers Coalition may be more accurate; their 2022 survey of current, aspiring, and past farmers under age forty found that more than 20 percent identify as a race other than white and almost 64 percent are female, nonbinary, or LGBTQ.[41] The results of both surveys show that America still has a long way to go in acknowledging agricultural racism and boosting diversity, even within the regenerative community.[42]

Even so, the changing demographics may foreshadow a return

to the original diversity of land stewardship from the 1400s and earlier, and to the original equality between men and women as related to food production. Women played a central role in Indigenous foodways and ecology. Native societies tended to be far more egalitarian, with women holding leadership positions and enjoying mutual respect with men.[43] In this way, "new" is not entirely accurate to describe Kelsey, since she is part of a long legacy of empowered Indigenous women. "I like to claim to be of the 125th generation to steward the land on the Great Plains, being a tribal descendant of the Lakota Nation," Kelsey says. "I come from a long line of land and community stewards on both sides of my family."

She highlights inspirational women family members, such as her mother, Vicki, who serves her community and family in too many ways to properly capture here, and who uplifted the ranch financially, physically, and emotionally during her many years there. Kelsey also cites her aunt Collette, a talented horsewoman; her aunt Lisa, who tragically passed away in an ATV accident while checking cattle; her aunt Lori, whose love is in feeding people, the centerpiece for most of the ranch's gatherings; and her late grandmother Regina, who Kelsey says was "the essence of the ranch" and "what bound the family together."

"I feel like I learned a lot from watching Granny and my mom interact growing up on the ranch. I apply these teachings every single day in being a woman with responsibility on the ranch," Kelsey reflected. "And they're just the women I was exposed to. I know the farther back we go, the more amazing female leaders we find." Like Kelsey's great-grandmother Babe (Claymore) Ducheneaux, who spearheaded the campaign to incorporate the Lakota word "oahe," meaning foundation or something to stand upon, into the name for the lake Kelsey and I overlook as we talk. "I wouldn't be as effective at doing what I do had it not been for what these women have taught me. But most of them never once called themselves a rancher, not even Granny, who lived on the ranch until her late seventies, when she passed away. She would have definitely called herself a rancher's wife."

Even though Kelsey sees "rancher's wife" as empowering terminology based on how her grandmothers lived the role, she also calls herself "rancher," and not just because she's formally educated as one. Kelsey is an equal decision-making and financial partner with her family at the DX Ranch. She's a respected leader, sharing regenerative agriculture information with her community and connecting other tribal producers with resources. Taken together, these traits make her very different from many women working within industrial agriculture.

Kelsey is also finding opportunities to take land stewardship a step further: in 2017 she launched DX Beef, the ranch's direct-to-consumer beef business. "Quite frankly, I don't feel like just managing the grass is good enough, because it's not helping to heal the food system. [Ranching] is really, really gratifying work, but it can be very disheartening the day that you drop your cattle off at the sale barn to be sold to the industry; you don't know who, you don't know where, and you don't know how they'll be treated," she says. "And then it's even more discouraging to drive home from the sale barn on your first sale day as a cattle producer, and you just drive by home after home facing food insecurity. That's where DX Beef started, was wanting to give my cattle the just life that I thought they deserved. They deserve to feed local people, and I wanted to be able to feel good about what I was raising cattle for. I wanted to know that I felt good eating the products that I was raising, and that my community members would be a little more connected with their food source."

Through the DX Beef arm of the ranch, Kelsey markets grass-fed, climate-friendly, locally processed animals directly to consumers via farmers' markets in Eagle Butte and through an online store. Customers can buy beef by the pound or in bulk. She knows many of her clients personally. Offering the Lakota community access to highly nutritious, affordable meat—a historic dietary staple for Plains tribes, as she reminds me—is a key part of her regenerative outlook, as is her dedication to restoring the consumer-food relationship. "It's not my goal for the whole world to be eating DX Beef; it's my goal for the whole world to be knowledgeable about a connection with their food and knowledgeable about where their food comes from," she says. Per year, the ranch processes forty-eight animals for direct marketing via DX Beef and sells around two hundred calves via live auction at the local sale barn. Operating this way means the DX Ranch contributes to building an alternative, regenerative food system for the local and regional community. And Kelsey seems hopeful that there will one day be a regenerative beef supply chain—one without feedlots and consolidated slaughterhouses, a system where producers are supported with fair prices and financing—into which she can readily market her live cattle.

The diversity and expertise that people like Kelsey bring to the table strengthen the food system and make the creation of a regenerative supply chain more likely, says Kathryn Brasier of Penn State. That is because diversifying the producer population means more ideas for adapting agriculture and the food system to a changing climate—and the world needs all the help it can get. How women and men deploy regenerative practices may not be radically different, but "how [women] learn [agriculture], how they take that and become creative with it, will be different based on the kinds of experiences they've had, the socialization they've grown up with, the networks and relationships which they are part of," Brasier says.

"All of that is a little bit different based on who you are, and that's also by race, ethnicity, gender, sexuality. The more creativity we can have, the more likely we are to come up with solutions to some of these tough problems."

The stakes are incredibly high. Industrial food is increasingly risky in a changed climate, but we have an alternative: we can create a genuinely regenerative food system, one that not only functions sustainably and produces nutrient-rich food, but also harnesses the perspectives, creativity, and talents of the whole population, of people like Kelsey. Disturbance, I am learning, may be the key to accomplishing all that. Disturbance is nature's way. From appropriate disturbance comes new life and lasting strength that make those disturbances less daunting. If we invite healthy disturbance into the food system, then we will be on the path to similar results—and achieve the resilience it will take to thrive on a hotter, more erratic planet.

2

Momentum

Carrie Martin and Erin Martin, Footprints in the Garden

Momentum: the traveling pattern exhibited by a wave, combining both oscillatory motion and forward motion.

That unforgettable line from the song "Gonna Make You Sweat (Everybody Dance Now)" blasts from a nearby speaker as I park my car in front of a vegetable-filled high tunnel and step outside. I've just arrived at Footprints in the Garden, a fifty-acre diversified vegetable and fruit farm co-operated by mother-daughter duo Carrie Martin and Erin Martin, located about an hour's drive southeast of Raleigh near Mount Olive, North Carolina. The overcast, blustery morning weather contrasts with the song's iconic dance beat, but that's the point. Carrie and Erin have fun farming on this land, and they want others to catch the spirit, too.

Carrie, fifty-seven, walks over from a field that currently contains kohlrabi, celery, Brussels sprouts, and broccoli. She wears a wide-brimmed black hat and brown leather work boots with black rubber soles. She greets me with a hug and what I quickly learn is her characteristic smile and optimism. "We try to keep things upbeat," she says over the music. "We like to have a good time around here." Carrie's positive outlook is one reason I am at her farm, and a big reason *she* is here. Despite countless challenges and a career path that at times felt uncertain, her determination to stay optimistic and find solutions ultimately led her to this land.

Carrie comes from a long line of Black farmers. But in the 1970s her parents moved into town to pursue better-paying jobs in health care and industrial engineering. Still, the family maintained a connection to the land. "Every year we had a garden," Carrie says. "We killed chickens. During family events, we barbecued hogs, had pig pickin's, all those good things." In high school Carrie met her future husband, Tim Martin, and the two attended college, she at Shaw University and he at Chowan University. Carrie worked in hotel management, then banking, and then pursued a master's degree in business from North Carolina State University. She and Tim bought a home in Raleigh and had their two children, Wesley and Erin. Motherhood was especially challenging when Tim, who served in the U.S. Army, was called overseas for long periods, like during Desert Storm. "It was hard because it was like being a single mom," she recalls. "He of course was doing what he had to do because it was during conflict. A lot of times it wasn't peaceful, campaigns where they were."

Carrie's professional life took more turns as her children grew up. She served as a social worker, then worked for ten years in the communications department in the College of Agriculture and Life Sciences at NC State, where she often supported the Cooperative Extension Service and became familiar with state and federal agriculture programs. She worked in the foster care system and in employment services and tended to her parents until they passed away. "I'm very proud of those things," Carrie says. "Just being able to be a community-supportive person throughout most of my careers, trying to help people advance and get educated, whether it's finance or being able to maintain your life." Then Carrie lost the job in employment services. So she did what she had counseled other jobseekers to do, which is to conduct a life assessment of the skills, assets, and interests she might harness to make a career change. One family asset stood out: the farmland in Mount Olive that she and Tim had inherited from his parents.

Carrie realized she was uniquely positioned to utilize that land for agriculture—a revelation enabled by her positive thinking. She came from generations of farmers. She understood how to access government agriculture programs and traverse the legalities of landownership. She could commute from Raleigh to Mount Olive because her children were older. She was passionate about health and community service. And she had a diverse professional and educational background that would lend itself to innovative farming and business ownership. "It led me right back here because of the finance, and the banking, and the hospitality, and working with employment, being able to put out job announcements and things like that. We do all of those things right here on the farm," Carrie says. She and Tim launched Footprints in the Garden in 2012, with Carrie as the primary farmer.

Carrie transitioned out of the USDA's beginning producer category (for people farming ten years or less) a year before I spoke with her in 2023. Though not technically a new farmer anymore, she's still part of the wave of women moving into agriculture over the last few decades, some traversing unlikely paths to get there. And like so many others in the movement, Carrie brings an ethic of environmental and social care to her work. She chose the name Footprints in the Garden as a nod to the marks we all leave on the Earth, and how connecting with the land in deep, authentic ways helps a person leave a positive footprint. "Farming is always associated with a lot of work, and it is. But it's also a place of tranquility and peace," she says. Erin shares her mother's vision of the farm as a site of healing. "I want it to be a place where people feel safe, people feel it's therapeutic. When things are going on, they can come to our land and decompress, whether it's from abuse, or PTSD from the army, or postpartum depression," Erin says. "I want our land to come to a person's mind when they're going through something."

Regenerative agriculture is one way to leave that positive footprint. And to Carrie and Erin, the idea of positive footprints is nothing novel; it is a continuation of agricultural philosophies passed down in their family that prioritize ecological reciprocity, responsible land management, and human health. "I'm doing what my forefathers and foremothers did. We come up every year with fancy words like regenerative, and organic, and all of those things. We're doing what our grandparents and great-grandparents taught us to do. And that's just to care for the land, to make sure you protect your family by not using the heavy pesticides, the heavy fertilizers," Carrie says.

As seen in Chapter 1, modern iterations of regenerative agriculture often draw from time-tested BIPOC wisdom. But regenerative agriculture's Afro-American origins are even less acknowledged than its Indigenous, Latin, and Asian roots. This denial is part of white America's long history (and continued habit in many circles) of erasing Black contributions to everything from music and cuisine to science and politics, erasures fueled by notions of white supremacy stretching back to the colonial era.[1] In truth, Africans developed regenerative forms of agriculture and land management well before white invaders arrived.

"African Indigenous agriculture was as diverse as the continent itself," writes Liz Carlisle in *Healing Grounds: Climate, Justice, and the Deep Roots of Regenerative Farming*. African farmers managed incredibly diverse and productive savanna farms and practiced layered agroforestry in the tropics to create food forests.[2] When colonizers trafficked them across the Atlantic, Africans carried with them their agricultural knowledge—and seeds for popular foods of today like black-eyed peas, okra, watermelons, and sweet potatoes—and deployed their seeds and skills in dooryard gardens while enslaved.[3] White slave owners often took notice of and exploited enslaved people's agricultural talents privately, while publicly dismissing them as ignorant. For instance, West Africans introduced rice, a crop cultivated primarily by women, using what we today call regenerative methods, such as rotating rice production with livestock grazing. Knowing West African women were highly skilled rice growers, white plantation owners in coastal South Carolina, Georgia, and Florida routinely paid more for those women and converted their expertise and labor into profit.[4]

More often, though, white farmers practiced extractive agriculture and forced enslaved people to participate. As a result of their destructive methods, white farmers were already abandoning spent fields in favor of fresh holdings to the west by the early to mid-1600s.[5] Three main factors contributed to their disregard for the land. First, most European immigrants simply were not good stewards. European farmland had been degraded by continuous cropping and tillage well before American colonization; in fact, soil exhaustion helped motivate European colonialism in general.[6] Second, seemingly limitless virgin land to the west discouraged farmers from conserving soil already under cultivation in the East and South. George Washington wrote extensively about the "unproductive" and "ruinous" agriculture going on in early America, as did Thomas Jefferson.[7]

Third, the chattel slavery system tied plantations to single-crop, extractive farming that necessitated more and more land.[8] From the 1700s until the Civil War, Southern plantation owners used slave labor to grow monoculture tobacco, cotton, and other cash crops without crop rotation, cover crops, or livestock integration, which quickly exhausted the soil. Plantation owners did not want slaves, whom they saw as interchangeable cogs in the agricultural machine, to "waste" time by sustainably tending diverse forms of crops and livestock. To slaveholders, the repeated and psychologically simple (though physically brutal) tasks of monoculture cropping better suited slave labor. When the land became too unproductive, slave owners cleared new fields or sold their plantations and moved west or deeper south.[9]

Slave owners eventually ran out of arable acres as more farmland fell under ownership and more states banned slavery. That is why slaveholders fought so fervently to establish slavery in new U.S. states and hoped to colonize the Caribbean and Central and South America—they were running out of land. When neither effort worked, slaveholders revolted against the United States government to protect their own interests. Soil depletion driven by the South's extractive, slave-based agricultural system was a major cause of the Civil War.[10]

Human and soil exploitation continued after the war, despite early efforts to prevent that outcome. Black leaders understood the economic and social power of landownership, and they told Abraham Lincoln's generals that their people wanted tillable farmland to make their own living, separate from white Southerners, who they feared would harbor hate well into the future. On January 16, 1865, the Lincoln administration's General William T. Sherman ordered the distribution of four hundred thousand acres of former Confederate land along the southeastern coastline to newly freed Blacks. Each family would receive roughly forty acres; Sherman later granted the army permission to lend mules to the families as well. By June, about forty thousand freed Blacks were settled on their land in self-governing communities. But meanwhile, the unthinkable happened: Lincoln was assassinated in April 1865. His successor, Southern sympathizer Andrew Johnson, annulled the land grant order in the fall and gave the land back "to the very people who had declared war on the United States of America," as historian Henry Louis Gates Jr. puts it.[11]

Atoning completely for the unforgivable sin of slavery is impossible, but America at least had a chance to give a mostly agricultural population of about 3.9 million people, people kept uneducated and financially under-resourced by law, *people whose enslavement the nation had institutionalized*, a fighting chance to thrive. Instead, the nation pretended it owed former slaves nothing. "Try to imagine how profoundly different the history of race relations in the United States would have been had this policy been implemented and enforced; had the former slaves actually had access to the ownership of land, of property; if they had had a chance to be self-sufficient economically, to build, accrue and pass on *wealth*," Gates writes. "After all, one of the principal promises of America was the possibility of average people being able to own *land*, and all that such ownership entailed. As we know all too well, this promise was not to be realized for the overwhelming majority of the nation's former slaves."[12]

Without access to land, and in the face of overwhelming white animosity in the Jim Crow era that followed, many Black people intent on farming took the only opportunity on the table: sharecropping. Of the formerly enslaved, the majority had labored in agriculture; farming was what they knew, and a profession open to them when most others were closed.[13] In this system, sharecroppers "rented" land to grow crops, and they gave the landlord, typically a white landowner, a portion of their harvest or proceeds as payment. The bigger the crop, the bigger the profit for both, or so the thinking went. But only cash crops like cotton, tobacco, and rice made sense in a skewed sharecropping system that demanded constant production to achieve even meager economic survival. Landlords could dictate what crops tenants grew and how, and remove tenants who disobeyed, all of which prevented sharecroppers from prioritizing sustainability. "When tenants didn't control their land and could be evicted at any time—or kicked off their land on trumped-up charges—they had little incentive to improve the soil," a *Smithsonian Magazine* writer explains.[14] Sharecroppers often had little choice but to continue extractive plantation-style agriculture.

On top of this, many landlords sold or rented farm necessities from on-farm stores at exorbitant prices, sometimes even mandating that tenants purchase from them on credit at high interest rates. This created a cycle of perpetual debt, and laws in some jurisdictions forbade indebted sharecroppers from leaving the land. To make matters worse, Southern soils were exhausted from decades of plantation-style cropping, which depressed yields and profits. The sharecropper system exploited white and Black farmers alike and has been called wage slavery and slavery by another name.[15] Keep in mind, though, that Black farmers were denied institutional credit and aid programs available to white sharecroppers, and they came to sharecropping with virtually no cash or assets.

Black thought leaders spearheaded efforts to correct American agriculture—and once again, the white agricultural world mostly ignored them. The brilliance of George Washington Carver's work at Tuskegee University's agricultural school in the early 1900s cannot be overstated. He showed farmers how to incorporate nitrogen- and food-producing crops like peanuts, cowpeas, and sweet potatoes into cash crop rotations, which increased both soil fertility and food security. He famously developed hundreds of ways to market surplus peanuts as an added revenue stream. Carver extolled farmers to forgo store-bought fertilizer and chemical pesticides in favor of compost, animal manure, and natural pest control for environmental and financial reasons. He encouraged them to grow their own vegetables, practice agroforestry, and employ permaculture—true sustainability.

Carver's goals were twofold: liberate farmers from the yoke of industrial agriculture so they could achieve food and economic sovereignty, and make farming a holistic, environmentally sustainable act that works in tandem with nature.[16] During his life Carver did receive recognition for his ecological approach to farming, but today credit for inventing regenerative agriculture largely goes to white people, like J. I. Rodale, Aldo Leopold, and Wendell Berry. Same for Booker T. Whatley, another Black regenerative agriculture advocate from Carver's era who often goes uncredited. Whatley also invented the CSA model of direct marketing to advance social and economic justice for small-scale Black farmers, a model now used by farmers around the world.[17]

Even lately, praise for regenerative agriculture innovation tends to go to white farmers, scientists, and activists, like Allan Savory, David Montgomery, Gabe Brown, Nicole Masters, and Robert Rodale. The 2020 film *Kiss the Ground* is an especially obvious example of how the movement prioritizes white voices. But countless BIPOC-led organizations and farms are pushing regenerative agriculture forward in groundbreaking ways, with women at the forefront. For instance, Angela Dawson founded 40 Acre Cooperative in Minnesota to empower Black farmers and ranchers with mentorship, training, access to markets, and financial assistance through the co-op's collective buying power for items like seeds and farm equipment. Dr. Cindy Ayers Elliot left Wall Street to start a regenerative diversified produce and livestock farm, with the goal of combating Mississippi's high obesity and diabetes rates by channeling fresh, healthy food into the Jackson community. She also trains new farmers, especially young people of color. Clarenda "Cee" Stanley operates Green Heffa Farms, a North Carolina organic regenerative hemp farm that works to expand access to agriculture to underserved and under-represented farmers. Green Heffa Farms is also a certified B Corporation, the first farm with a Black female CEO to earn that status. A few more of the many, many other notable BIWOC leaders include A-dae Romero-Briones, Winona LaDuke, Karen Washington, Oliva Watkins, Leah Penniman, Mai Nguyen, Lyla June Johnston, and Aidee Guzman.

Situated in this long tradition of Black regenerative agriculture are Carrie and Erin. Their farm reflects Carver's vision of sustainability, employs Whatley's CSA model, and incorporates the diversity and sustainability of past and present Black agriculture. In addition to eliminating or minimizing sprays, they rely on a number of other regenerative practices and tools, such as no and low tillage, companion planting, polyculture cropping, and natural, organic sprays for pests or plant diseases they can't otherwise manage. They enrich soil with farm-produced compost, potash (a naturally occurring, non–fossil fuel–derived fertilizer), and fish meal. They install portable drip tape each season to deliver liquid compost tea and water (though the land rarely needs irrigation thanks to good soil health). That drip tape allows targeted fertilizer and water applications, which reduces runoff and overapplication. Carrie and Erin control weeds by hand, through intercropping, and by covering raised beds with limited plastic that also conserves water. They grow all manner of vegetables—leafy greens, root vegetables, asparagus, tomatoes, legumes, celery, herbs, way more plants than I can list here—as well as fruits like watermelons, blueberries, raspberries, elderberries, and grapes.

"We try to do niche market kind of things, like we don't just grow a regular beet. It'll be something like an orange beet, or it'll be the kohlrabi, which is purple or white, or three or four kinds of radishes, which will be like an Easter egg radish or black radish or a watermelon radish," Carrie says. "We're just trying to do nontraditional-type things to get folks interested in and learning about different vegetables that are available for their health, and to experiment and see if they really like it. We really believe that food is the key to good health." To diversify even further and attract pollinators, Carrie and Erin recently put in an orchard of around seventy-five peach, pear, fig, persimmon, papaw, and plum trees. They plant crops within the orchard, a practice called agroforestry, to build soil health, control weeds, and boost biodiversity. A forest stands on the farm's far end, which provides wildlife habitat, cools the air in summer, protects crops from wind damage, and will serve as an income stream when the forest matures enough to practice sustainable logging.

Carrie and Erin rotate where they plant annual crops each year, with the remaining fields devoted to cover crops like the buckwheat I see nearby. Cover crops, which are crops planted before, with, or after cash crops, are an essential regenerative practice. For example, row crop farmers can plant a rye/vetch mix after a corn crop. Or they might interseed corn with sweet clover, buckwheat, or cowpeas so everything grows together. Cover crops keep a living root in the soil year-round, which sustains soil biology. They provide diversity for the above- and belowground ecosystems. They also keep soil covered to control weeds, stop erosion, and prevent moisture loss, a critical benefit in a warming world.

"Soils hit 100 degrees, and everything in there dies. And if it's wet, it's going to be worse because the water inside it is going to boil," explains Dr. Jill Clapperton, principal scientist and CEO of

Rhizoterra Inc. and the founder of the online Global Food & Farm Community. "This is why we need insulation on the ground. This is why we're doing what we're doing with regenerative and trying to keep mulches on the ground, trying to keep plants growing all the time. When it's 115 outside, if you have plants growing, it can be 80 degrees under the canopy. As long as it's 80 degrees under the canopy, that plant will keep sweating, keep transpiring, which means it will keep pumping up nutrients, and it will keep growing and it will survive the heat stress. It will retain its biological components and maintain the infrastructure that's belowground."

Cover crops help soil stay alive through the winter as well. "You don't have to just have bare ground in the winter," Clapperton continues. "Under a good snow cover, as long as the soil isn't frozen, the microbiology is working. Microarthropods are working; they're all working." Again, because of climate change, a cover crop's winter benefits become amplified. Winter moisture manifests more often than it once did as rain, which runs off of bare soil. Snow that does materialize is increasingly likely to fall in extreme bursts, and bare soil can't hold that either. Warmer winters also expose bare soil to higher temperatures over longer durations, further reducing biological life.

Erin is still on her way from Raleigh, so Carrie leads me on a walking tour of the farm. The first stop is a brick farmhouse just off the paved road, the house Carrie's in-laws built and where Tim grew up. Carrie shuts off the speaker on the front step so we can hear each other better. The house functions as storage for produce and farm tools and as a staging area for packing CSA boxes. That's soon to change, though, because in the coming months Carrie and Erin are converting an adjacent building into a cold storage area with a CoolBot walk-in cooler. Erin won grants from Rural

Advancement Foundation International-USA and the North Carolina Tobacco Trust Fund for the remodel. Majestic pecan trees, planted by Tim's parents, stand watch at the house's front and rear. These trees provide another revenue stream through the sale of pecans and pecan-based products. Carrie points out one of two Martin family cemeteries about a hundred yards away. Living relatives are nearby, too. Three other Martin families farm the surrounding land, which has passed from generation to generation. "I'm so thankful for Tim's ancestors, because they had a vision of what they wanted. And it was for us to stay together and work together," Carrie says.

The farm is currently in transition as winter and early spring crops wind down and late spring and summer crops replace them. Determining when to plant is trickier than it used to be as climate change ushers in more frequent extreme weather events, like unseasonal temperature swings. "North Carolina weather, because of the climate changing, we just don't really know [when the last frost will be]. But we're going to try to plant somewhere the first week of May to put more of our spring and summer crops in," Carrie tells me as she points out where asparagus beds are established and where the watermelons and sunflowers will go. The sunflowers are a new addition, part of Erin's larger effort to scale up agrotourism. "She's also a photographer, so her vision is to have a sunflower field so that she can use her photography skills and invite other photographers and families out to take pictures," Carrie explains. Carrie shows me two more agrotourism elements, the farm's meditation garden and the spot where she and Erin hope to construct a building for hosting community events. The building would also replace the old farmhouse as an aggregation center, which would give them space to bring in even more produce and value-added items from other farmers for their CSA boxes.

Then we enter the pièce de résistance: the high tunnel, which allows Carrie and Erin to extend the growing season. The sugar snap peas, leeks, celery, Swiss chard, carrots, tomatoes, herbs, and beets growing now represent the last of the spring harvest. Soon the tunnel will be brimming with summer crops. Our final stop is the greenhouse, where Carrie and Erin start seedlings. They buy additional seedlings from a local nursery, yet another way small farms like Footprints in the Garden support rural economies. Looking around, I see Carrie's vision for a positive footprint taking many forms: helping the area's farms and food businesses market products, stewarding the land, minimizing carbon emissions, welcoming others to the farm, and providing fresh, nutritious food.

As the regenerative agriculture movement grows, I can also envision a future where more farms like Footprints in the Garden help regenerate rural communities across the nation—in other words, a future more aligned with Carrie's optimistic outlook. "We've maintained and survived, and we try to keep it as lively and as positive as we can," she says. "Our place is open. We want everyone to be valued, whatever their beliefs are, their cultures are, whatever their sexual orientation is. That's up to them. We're wide open here."

Erin arrives with her daughter, Elana, a sweet-natured fifteen-month-old in a purple hoodie with curly hair and a charming smile. Erin often brings Elana to the farm, as her brother Wesley does with his children, Julian and Jett. Elana is curious about everything; she especially likes patting dirt into seed trays, Carrie told me earlier. She and Erin joke that Elana is the "farm manager" since her moods can dictate how the day goes. Erin plans to teach Elana the ins and outs of growing food, an empowering and necessary skill. "There's a lot of people that don't know how to grow their own food," Erin says. "There's a lot of people that don't even know how to put the seed into the ground, and you have to keep watering it, and have the right temperature, and know what season to grow in. It's important for our children to know that."

The knowledge gap is not always a function of people not caring, Erin points out. Often they lack exposure to agriculture and nutrition information, or require more resources, time, and connections to pursue those interests. But Erin sees a shift in her generation toward a greater interest in food, health, and the environment, a shift driven by multiple factors. "I think a lot of people are interested [in farming], definitely interested in their health, and I think 2020 gave us that scare that we needed to bring us back to the land. I think it's definitely something that people think about," she says. "As I get older, my group is having children, so that's a realization check for them as well, too, to be able to put fresh, affordable food in front of their children."

Erin admits that as a child she did not always appreciate the farm or why knowing how to grow food is important. She credits Carrie with cultivating curiosity and modeling how to keep learning. "I have to give her props on that because she keeps us up to date on those different things. We were going to workshops when I was younger to learn about these different things, but I wasn't paying attention because I'm ten, twelve years old, just like, 'When are we going home?' I wasn't paying attention to that," Erin recalls. "But now I am taking notes, I'm making sure that I know exactly how to do these things or how to install and repair the drip tape, and I'm able to pass those down to my daughter." Today Carrie and Erin attend educational events together. They also lead them. When they install the CoolBot next month, for instance, they will turn the day into a workshop so other farmers can see the process.

One reason for Footprints in the Garden's success is this combination of Carrie's experience and Erin's Gen Z perspective. Like many of today's young and beginning farmers, Erin leverages social media to market products and promote agrotourism enterprises. She creates content that keeps customers engaged and, ideally, eager to buy more. In addition to CSA boxes, Carrie and Erin market through North Carolina food hubs in Wilmington and Durham. They also sell at the Black Farmers' Market that alternates between Raleigh and Durham.

Communication is vital for marketing, Erin explains. At the markets especially, she says, it's about "just talking to people, not being scared to talk to people about what you do, how you want to display things. You can't be scared to talk to people about what you want to do, what you believe in, because how are they going to know if you don't talk about it? It's getting out here and meeting new people." Social media and one-on-one conversations also lower the veil on agricultural life, allowing the public to see what farming and ranching look like—that farmers are not uneducated country hicks who look like they work in the fields all day. They have pride in their work and how they look. And when producers use regenerative practices, there is nothing to hide.

Social media is only the beginning. Erin ignited the farm's push into agrotourism, like hosting farm-to-table dinners—a vision she had even before joining the farm. "That's something I really saw a long time ago because I've always liked cooking. I've always been the cook of our family, whether it was cooking something untraditional or traditional and trying to put a spin on it," she says. "Seeing a vegetable go from seed to full flourishing and then turning it into a meal that feeds your family and friends is something that I always prefer doing." This season she has plans for a kids' sunflower maze and a big Juneteenth celebration. Such onsite business is easier now that, thanks to Erin, the farm accepts credit and debit card payments. In addition to agrotourism, Erin introduced farm-derived value-added products, like her elderberry syrup and pepper jelly, to diversify the farm's revenue even more.

As we chat, Elana scoops dirt from a bucket with a trowel. She laughs and gleefully shows us her dirt-covered hands. The soil she plays with is more than a life-giving resource—it represents a legacy first seeded by Erin's great-great-grandfather, Harry. Harry and his siblings were born into slavery in Sampson County, North Carolina. In the 1860s, when Harry was sixteen, he came upon the slave master beating his sister. To save her, he struck the master in the head and ran for his life. The master pursued with dogs, but Harry's sister hid him in the hollow of an oak tree. Traveling by night, Harry escaped to the neighboring county, Wayne County, and not long after joined the 135th United States Colored Troop infantry division.

When the Civil War ended, Harry returned to Wayne County and in 1883 purchased land for himself and his five children. The land passed from generation to generation for more than a century, with family members purchasing additional acreages through the years and keeping their family heritage alive in different ways. Erin respects Harry's tenacity and generosity, which create positive ripple effects in her family's lives to this day. "He had enough education and literacy to write a deed up and preserve it, and register the deed with Wayne County. We were able to obtain [Century Farm] status in 2022," she says. "But that just is mind-boggling because a lot of slaves didn't know how to read, they didn't know how to write. The fact that he went on to be able to write a deed up so that we could obtain that status today is amazing."

That the Martin family still owns their land is remarkable given the systematic assault on Black land rights over the past 150 years.

Black Americans owned between 16 and 19 million acres of farmland in 1910, the peak of Black farmland ownership.[18] Today, Black farmers own less than 3 million of the 4.7 million farmland acres they operate, according to FarmAid.[19] The stark decrease is in large part the result of centuries of structural discrimination that separated Black Americans from their land.[20] One example of such institutional racism is heirs' property laws.[21] If a landowner dies without a will stating who should inherit the land—referred to as "dying intestate"—then the land becomes heirs' property, or property owned in common by family members.

One can imagine the legal nightmares of common ownership, especially in a big family. Because these heirs lack clear title to the property, they cannot use it as collateral for loans or collect various forms of federal and state assistance. They are jointly responsible for taxes and maintenance. Heirs may die without wills like their relatives did, further fragmenting ownership. Heirs may not live near or even know one another. As Thomas W. Mitchell, legal scholar and director of Boston College School of Law's Initiative on Land, Housing & Property Rights, notes, "Because of these characteristics of heir property, economic development of a significant proportion of land owned by African Americans has been stifled. Owners have difficulty obtaining financing and coowners may not be able to agree on the most appropriate use of the land."[22] In short, heirs' property cannot function as a wealth-building asset.

Studies reveal that at least half, likely more, of rural Black landowners in the South do not have wills. The result is that heirs' property is prevalent in rural Black communities compared to white communities. Mitchell argues that the reasons behind the lack of wills are not well understood but seem to center on access. We know that the legal system and law as a profession were largely off-limits for Black people in the late 1800s and early 1900s, so many never gained titles for purchased or inherited tracts. Many could not seek wills either. Although legal access improved in the post–civil rights era, many rural Black families could not and still cannot afford legal services. Mitchell also points to studies showing that many rural Black landowners do not fully understand the heirs' property laws, which further suggests a lack of access to legal professionals and the knowledge they provide.

The heirs' property system seems designed to separate families from their land. In that system, white judges often decide the fate of Black-owned property—and that can lead to discrimination. For instance, when even one heir asks a court to dissolve the joint tenancy, judges can order partitions in kind that physically divide land among family members and grant individual titles. But courts typically order partition sales instead, which force the sale of all the land and divide the proceeds among the heirs. As one lawyer wrote, "Scholars attribute this preference [for partition sales] to a variety of motivations, ranging from implicit or explicit racial bias to the application of economic analyses to merely preferring the simplest alternative."[23] Whatever the motivation, the result is usually the same, and it is hard to imagine that judges don't know it: Black families lose their land in the end. Heirs could buy the land back at auction, but white buyers with more cash and borrowing leverage typically outbid them and other would-be buyers of color.

Heirs' property laws still dispossess Black families of their land. That is why Carrie serves as the director of agriculture and conservation at Black Family Land Trust, one of the nation's only conservation land trusts dedicated to the preservation and protection of African American and other historically underserved landowner assets. People like Carrie and organizations like Black Family Land Trust are working to close the legal knowledge gap and motivate rural Black landowners to utilize the system to protect their land

and legacy. "I learn from those farmers that I help, and I help them by bringing them more of the technical type of things, like the easements, helping them try to avoid heirs' property, but be able to exist if they are in a current heirs' property situation," she explains. "To educate the youth about the power of being landowners."

Heirs' property laws are not the only form of structural discrimination. Black farmers often missed out on institutional agricultural assistance, which made keeping their farms harder than it otherwise might have been. White people controlled, and usually still do, the government agencies, land grant institutions, nonprofits, agribusinesses, and other farm-related entities that provide aid, support, education, and economic opportunities. That power fueled both deliberate and passive discrimination against Black farmers, as well as preferential treatment for white farmers. For instance, white administrators of New Deal agendas denied loans and sharecropping work to Black farmers.[24] USDA bureaucrats tailored aid programs for large-scale commodity farms while bypassing smaller farms growing more labor-intensive, high-value crops like fruits and vegetables—the type of farms BIPOC and women operators tend to run.[25]

Until the Voting Rights Act of 1965, Black farmers faced Jim Crow bureaucratic restrictions, bullying, physical violence, economic punishment, and other barriers in voting for representatives who would advance their interests. In addition, "Lynchings, police brutality, and other forms of intimidation were sometimes used to dispossess black farmers, and even when land wasn't a motivation for such actions, much of the violence left land without an owner," writes journalist Vann R. Newkirk II in *The Atlantic*.[26] Such harassment was particularly egregious in the South, where most Black farmers lived.

Even after the civil rights movement, the USDA continued to systematically discriminate against Black farmers, as evidenced in numerous assessments of its practices and by the successful discrimination lawsuit Black farmers settled in 2010. Indigenous, Hispanic, and female producers went on to file and win similar lawsuits. Given this history, it's no wonder distrust lingers between BIPOC farmers and the federal government.

Carrie tells me about the skepticism she faced from a white Farm Service Agency (FSA) official when applying for COVID-19 farmer relief funds. The problem: the agent claimed that the word "producer" in the aid program's guidelines applied to large-scale farmers only. She was essentially telling Carrie and Erin that they were not real farmers. "I had to bring them the statement that this is what the literature said, and that other counties are taking advantage of this and they're not interpreting that word producer as a big-name producer," Carrie says. "I was told, 'Okay, yes, we're going to put your paperwork through, but there's a possibility that you may be audited.' Needless to say, the next week or two later, I got a letter saying that I was being audited. So I had to gather all my paperwork, and luckily we do fairly well with our paperwork. I was able to send them the paperwork which they needed. I don't think they looked at it. I think they just didn't feel like I had the paperwork."

Carrie and Erin are not the only farmers of color to encounter this kind of suspicion. In 2020, the USDA accepted just 37 percent of Black farmer applications for a program that helps pay for land, equipment, and repairs. Meanwhile, the agency accepted 71 percent of white farmer applications.[27] For a separate COVID-19 relief program, less than 1 percent of funds went to BIPOC farmers, despite the fact that such farmers make up 5 percent of the total farming population.[28] Among young farmers, Black and

Indigenous producers are more likely than their white counterparts to report that local and state USDA officials ignore them or treat them in unwelcoming ways, and more likely to be denied federal assistance.[29] "I think for me, and I think for a lot of farm families, we just want to have fair access to programs. A lot of folks are being denied program benefits that they pay for with their tax dollars," Carrie says.

Unsurprisingly, the presence of Black farmers in the food system has declined because of mass dispossession and structural discrimination. In 1920, the percentage of Black farmers peaked at 14.2 percent of the farming population.[30] Today, more than 95 percent of U.S. farmers are white, while 1.3 percent are Black.[31] Researchers estimate that Black-owned farm loss was more than twice that of white-owned farm loss in the past half century.[32] White control of farmland contributed to the erosion of Black food cultures and agricultural practices. It further spurred migration of Black people to cities, lowered diversity in rural communities, and severed career paths for Black children. Then there are the economic consequences: Black families lost billions in generational wealth due to land loss, the equivalent of $326 billion in today's dollars to be exact.[33] What's more, under white farmers' watch, agriculture became industrial, all about dominance of the land rather than harmony with it, a corporatized race for profits at any ecological cost. I am hesitant to delve too far into what-if scenarios, but I do believe that if BIPOC, LGBTQ+, and female farmers had been fairly represented in agriculture, then food and farming would look radically different—more environmentally sustainable, more equitable for all involved, more nutrition focused, and more climate-friendly.

Despite all the obstacles, the number of Black primary farm operators grew from 33,371 to 45,508 between the 2012 and 2017 Censuses of Agriculture. The number of Hispanic and Indigenous operators jumped by even larger margins. As with the latest data on female operators, some of the BIPOC producer increase is the result of a new census format that allows respondents to list more people involved with farm decisions. But also as with female producers, the modified format accounts for only some of the growth. The survey often fails to count the types of farms many BIPOC producers run. Compared to most conventional farms, these under-counted farms tend to be smaller; nontraditional in crop type, business model, and location; and more centered on sustainability. However, the USDA's focus historically is large, industrial operations. "[The Censuses of Agriculture] tend to miss urban and they tend to miss small, [and] African American and Latinx farmers are going to be more in that space," says Kathryn Brasier, professor of rural sociology at Pennsylvania State University.

The damaged relationship between farmers of color and the USDA is another reason the census likely does not reflect the total number of BIPOC farmers, because those producers may not respond to or even receive a census form. In addition, non-USDA research by academics and farm advocacy groups tends to reflect the white farmer experience, creating a knowledge gap about BIPOC farmers. Looking at research on female growers in particular, scholars note that "while the research on women in agriculture in the US is of fundamental importance, too often, this work tends to suffer from either an oversampling of white women, discussions of research participation that do not include the racial demographics of participants (or makes them a footnote), or discussions of research findings that ignore the centrality of race to theorizing and understanding the experience of women in agriculture."[34]

Even if the data is incomplete, the fact remains that the number of BIPOC farmers is rising. Erin is one of those newcomers. She officially joined the farm in 2020 at age twenty-four, but she studied, worked, and advocated within the food system for years prior. In high school, for instance, she had to pick a career and research it as part of her senior high capstone project. Erin chose dietitian, a natural inclination given her love for cooking. Diving into the world of food and nutrition, she learned that southeast Raleigh, where her high school is located and not far from where the Martins live, is a food desert. The USDA defines food deserts as "regions of the country [that] often feature large proportions of households with low incomes, inadequate access to transportation, and a limited number of food retailers providing fresh produce and healthy groceries for affordable prices."[35] Urban and rural food deserts share some common demographic traits. Compared to other areas, they tend to have smaller overall populations but greater concentrations of minority groups.[36] The USDA has four ways to measure food access, but the common understanding is that an urban community is a food desert if the nearest grocery store is more than a mile away for the majority of residents. For a rural community, the distance is more than twenty miles.

Erin realized that the produce her mom grew could feed underserved families in her hometown community. She and her mom could be "food desert fighters," as she frequently puts it. "That's what led us to Raleigh City Farm," she explains, an organization she learned about through her capstone research. Raleigh City Farm is a nonprofit urban farm founded in 2011 on a formerly vacant one-acre lot in downtown Raleigh. Its mission is to reconnect city dwellers with healthy food production and nourish the diverse Raleigh community via a Pay-What-You-Can Farmstand and by donating about 50 percent of its produce to local nonprofits addressing food insecurity and access. At the time, the farm also ran a food hub that aggregated food from nearby small farms and sold it to local restaurants, hotels, and Raleigh residents.

Erin introduced Carrie to the farm's leadership, and soon Footprints in the Garden was marketing product through Raleigh City Farm. The partnership facilitated connections with other farmers and leaders in the local food scene, which inspired Carrie and Erin to experiment with new farming techniques and crops. Over the years Footprints in the Garden elevated its presence in the community; Beyoncé and Jay-Z even dined at a restaurant that featured their vegetables.

Meanwhile, Erin attended Wake Technical Community College, taking general classes that would allow her to transfer to a four-year university to obtain a bachelor of science degree in agriculture. But like the plans of so many other Americans, Erin's plans were derailed when COVID-19 hit in 2020 and she lost her job. She found herself asking a question similar to the one Carrie had faced about a decade prior: how to marry a diverse set of skills and work experience with a passion for social justice, especially in relation to food and health? The question took on greater urgency as food supplies tightened as a consequence of the pandemic; Erin's community needed the produce Footprints in the Garden was growing. Plus, she had a newborn baby to care for. And so as it was for Carrie, Erin's answer to the question of how to move forward was on the land.

Another way Erin's life echoes Carrie's is through service. While she plans to attend a local university in summer 2024, she also has her eye on becoming a registered forester to increase her knowledge of the land and provide services and information to landowners about forested land. The U.S. forestry sector desperately needs more people like Erin. There is a shortage of foresters generally, and less than 3 percent of foresters and conservation scientists are Black and even fewer are female.[37] Erin's commitment to sustainable forest management and regenerative agriculture even made her the winner of the inaugural Carolinas Leopold Conservation Award in 2023.

That's not all when it comes to service. Since 2022, Erin has worked as community engagement coordinator for Fertile Ground, a Raleigh food cooperative working to increase access to healthy, affordable food. Her role is to keep members engaged and up to date about Fertile Ground's efforts, a task informed by her farmwork. "I feel like that brought my whole world full circle. Because you have this community that you need to feed, you have this land, you need to start doing something with it, as well as feeding the community that you spent so much time in," she says. "Our people need that food. Our children need that food." The co-op plans to open a grocery store and community gathering space in southeast Raleigh in 2025; in fact, Erin participated in a site reveal event the day before my farm visit.

Erin's work with Fertile Ground, and Carrie's with Black Family Land Trust, are part of the multidimensional aspect of regenerative agriculture. Regenerative agriculture renews both urban and rural communities, and it nourishes both human bodies and the planet. It reaches back into history, honoring its diverse ancestors while welcoming new recruits. It remedies social inequities and repairs broken relationships. Such agriculture also builds soil fertility, ecological resources, and revenue possibilities for future generations. Unlike industrial agriculture, which degrades soil and too often leaves the next generation in debt, regenerative operations are better positioned to leave a legacy. Erin thinks about all this more now that Elana is in her life. "I hope that I can create something easier for my children. There's also aspects and parts of farming that have created a work ethic in me that I think is not in a lot of younger individuals. So maybe not too easy, but *easier*," she says. "I hope I can create something that will put reassurance in their mind to want them to come back to the land."

3

Direction

Susan Jaster, Lincoln University Extension

Direction: the course or line along which waves
propagate and travel.

By the time I exit I-70 for Oak Grove, Missouri, the storm is an ominous blue mass on the horizon. I've been glancing at the clouds in my rearview mirror for the last few hours. Now that Tornado Alley has shifted east and south—potentially a consequence of a warmer Gulf of Mexico—Missouri sees more tornadoes than it used to.[1] I'm hoping not to see one this afternoon.

I park at a bustling truck stop. Soon Susan Jaster, farm outreach worker with Lincoln University Cooperative Extension, pulls up in her pickup truck. I hop in and we cruise down narrow corn- and soybean-lined roads that Susan tells me flood regularly. The region experiences more intense rain events now, and the industrially farmed soils around us lack the organic matter needed to absorb the water. This area of Missouri is part of the Corn Belt, a section of the Midwest planted heavily to genetically modified corn-soy rotations. That corn is not what consumers see in the grocery store. Less than 1 percent of the corn grown in this country is sweet corn. Virtually all the corn U.S. farmers plant is dent corn, a variety inedible in raw form to humans but ideal for feeding to livestock. It's also perfect for converting into ethanol fuel (which is at least 24 percent more carbon-intensive to produce than gasoline, according to the U.S. Department of Energy and National Wildlife Federation[2]) and processing into unhealthy ingredients like high-fructose corn syrup.[3] Same with soybeans: the vast majority of U.S. farmers grow soy for livestock—by far the most common use—and human consumption that requires processing first.

Susan and I are heading to a farm owned by one of five area women (including Susan) who are using a Sustainable Agriculture Research and Education grant to analyze the relationship between grazing, wild forb and insect diversity, and soil health. The two-year study, in its final year when I visit in summer 2021, compares test and control plots within cover crop pastures located on each woman's farm. The test plots undergo rotational, high-intensity livestock grazing, while the control plots are not grazed. "Cover crop" typically means a temporary crop planted in rotation with cash crops. But because these women primarily raise livestock, their cover crop fields function more like pastures. They do not till or plant the land to cash crops at all. Instead, they seed the fields annually with a cover crop mix, and other plants naturally work their way in. Sometimes covers from previous years return. The women evaluate the plots after each round of livestock impact.

"We're looking at what population of plants we have in there, whether it's some of our cover crops or if it's wild forbs, or weeds. We're counting the insects that are on there, and we're also see-ing what relationship of predator to prey bugs that we have. We're going to try and see from beginning to end of the project if those insects change and those plant communities change," Susan tells me in an earlier conversation via Zoom. "We come in and do a Solvita test, which is the way to tell how many carbons you're sequestering, basically. We're also doing a microBIOMETER test, which tells you the ratio of fungus to bacteria. Those are all things we're trying to achieve, carbon sequestration and a good balance of bacteria to fungus, for regenerative practices. We're also doing a water infiltration test, which usually is a good indicator that your soil is getting better." The goal is to take a holistic view of regenerative practices—to assess the overall effect instead of just one or two outcomes. "Usually when people do this kind of research they get too microscopic," Susan says. "You need to look at more of the holistic picture because that's what regenerative practices are—fitting the whole together so it works properly."

Research is an important part of Susan's work, but her main job is assisting farmers via Lincoln University Cooperative Extension's Innovative Small Farmers' Outreach Program. The program's mission is to help small farmers and ranchers protect soil, water, and the environment while also boosting their efficiency and profit. As a farm outreach worker since 2009, Susan offers advice on strategies to lower input costs, improve farming skills and recordkeeping, increase yields/production, integrate new enterprises, find niche markets, hone marketing skills, and add value to harvests and products. The program aims to reach underserved and socially disadvantaged producers in particular, two groups the USDA, extension service, universities, and other agriculture entities have historically ignored. The program also has an urban farming arm that promotes urban agriculture and provides knowledge to those growers.

Susan works with farmers one-on-one, tailoring her advice to their environment and goals. She also speaks publicly and conducts informative workshops. She specializes in sharing innovative ideas related to soil health, clean water, grazing, pasture/field management, livestock, low- to no-chemical use, biological farming, and other regenerative agriculture priorities and practices. "I feel like in the last four or five years I have finally figured out that is really my point and purpose on this planet, to make sure that people know how to do this, so that at least the generations that are on the planet right now will have abundance in their final years and get their children to understand that this is possible," she says.

One reason Susan is so committed to regenerative agriculture, and so effective at helping farmers understand it, is that she is a farmer and rancher herself. She started in the dairy industry in the early 1980s on a large commercial dairy in Arizona. She continued in Missouri on a 150-cow dairy that she co-owned with her husband, Art. She cared for their three children during those years as well. Since 2010 Susan has raised American Blackbelly sheep, a hair breed that thrives under low-input grazing conditions and also yields high-quality horns and lean meat. Like the cattle at the DX Ranch, sheep play a starring role in regenerating the land. At the time of my visit, she has roughly one hundred head of sheep, and their home base is the ten acres of pasture and cover crop fields surrounding her house. Susan also rents additional grazing acres that she manages with regenerative practices.

Renting is a challenge, though, because leases are not guaranteed. Decision-making freedom, wealth accrual for future generations, and the ability to borrow against the land asset are several reasons producers prefer to own. Renting can also complicate conservation. The dominant narrative is that renters are extractive, putting little energy into conservation of land they may only control for short periods of time. But that is not always true, as people like Susan and other regenerative renters I've spoken with demonstrate. In some cases, tenants may want to farm or ranch regeneratively, but the landlord may discourage or forbid it, believing it will reduce profit. The situation might also be reversed, with landowners encouraging sustainable practices that tenants reject. What's needed is a system for matching like-minded landowners and tenants, as well as incentives for landlords to require sustainability and tenants to practice it willingly. Given that roughly 40 percent of U.S. farmland is rented,[4] centering landlord-tenant relationships around sustainability could make a substantial environmental impact.

And women could play a major role in doing that. *Women own almost half of the rented farmland in America*, accounting for 37 percent of landlords.[5] Research shows that both male and female nonoperating landlords care about conservation at roughly equal rates; however, women report less knowledge about and engagement with conservation practices and resources.[6] They often lack the confidence and know-how to dialogue with their tenants about regenerative management. To address this, organizations like American Farmland Trust and the Women, Food and Agriculture Network offer programming for female landowners focused on sustainability. "Particularly in parts of the Midwest, we have high rates of rented land, and much of that land is owned by women, widows or women who have inherited that land, but who are less actively managing it," says Gabrielle Roesch-McNally, Women for the Land director at American Farmland Trust. "The focus of our early programming was on engaging those women, primarily in conversations around conservation, acknowledging that if we want to get conservation and more resilient practices, regenerative practices, on the landscape, then we need to make sure women are part of that conversation on their rented acres."

Women landowners could also create pathways for young and diverse producers to access land. Land access is the number one barrier for new farmers, especially those of color, so we ought to build systems for connecting women landowners with young and BIPOC producers. The land-matching program at the Center for Rural Affairs offers a template for how we might match young, diverse farmers with women landowners who want to see the next generation thrive.

Susan and I arrive at Dawn Hoover's house, where Kelly Clark, Cathy Geary, and Sariah Hoover (Dawn's daughter) also await us. Aside from Cathy and Susan, who've farmed most of their lives, the other women fit the USDA definition of "beginning farmer." Dawn, sixty-seven, used to rent all her land to an industrial grain producer, but in the last five years she has operated more and more of it herself. She also manages part of a neighbor's acreage, with Sariah, forty, helping her. And Kelly, thirty-nine, started her diversified ranch almost six years ago with her husband. As the clouds darken, we march out to the test plots. "This area down here is only three years from conventional," Dawn says as we enter the cover crop pasture. "Each year it's doing better, but this year it's doing *a lot* better."

Susan stops at one of the thin yellow poles that serve as transects. "We've been inventorying a one-meter square off of these posts," she says. "We inventory by percentages. Whatever is in there, we write it all down. Like this is black medic right here. It's a legume, it looks like clover, but it's a little more veiny. It has all these little lines in it and it has a darker stem usually. It is trying to repair the soil because it needs calcium, it needs some nitrogen, so it's helping with that. But she's also got some other things in here, radishes, turnips, some wild forbs. Here's some wild oats."

"I had 32 percent fungal bacteria on this area," Dawn adds, pointing to one of the test areas. "The control plot over there, which you can look at and tell it's not doing as good, it was 8 percent." Fungal bacteria may sound nasty, but it's actually crucial to plant health: fungi cycle nutrients, fix nitrogen, produce beneficial hormones, control pathogens, protect against drought, stabilize organic matter, and decompose plant, animal, and other residue.[7]

Except for Cathy, these women manage land that others farmed industrially in the not-so-distant past, land with essentially no soil life left when they started. Regenerating such land takes time—but not as much as one might think. "My pasture out in front of my property, I've taken it back from corn and bean land, and it took every bit of three years to get it to where we had live biological activity in that soil," Susan tells me earlier. "It can happen quickly, but you have to be purposeful in your actions."

An example: Kelly recalls a parasite problem she faced when she first put livestock on her land. Because the ecosystem was so out of whack from decades of industrial farming, parasite numbers jumped high enough to kill animals. A nosy neighbor commented to Kelly's mother that he couldn't understand why Kelly and her husband were trying to run animals there. Couldn't they see the land was dead and needed agrochemicals to fix it? That's the wrong mentality, Kelly says. "Now I get to regenerate it and figure out how to utilize it," she says. "Susan saw it from when it started and through this grant and where we've been in changing it, and it's a huge difference. My loss of animals due to parasites has pretty much been nothing this year. This property isn't dead; it just needed some healing and it needed some time."

The nosy neighbor happened to be a man, and his response to regenerative practices is typical not just in rural Missouri, but in much of America's farm country. The women are quick to say that regenerative is never anti-male, nor is it woman-only (a sentiment I and other women featured in this book would heartily agree with). But they also acknowledge the pushback from male peers: derogatory comments, accusations of being crazy, dismissals of their work as "not real farming." I've heard that before, and so has Susan. Women and men alike who use regenerative organic practices often receive skepticism and even outright scorn.

Conservatives have branded regenerative agriculture as a liberal agenda, a model farmers ought to reject as antibusiness and anti-progress, a threat to their existence. Some farmers shun regenerative agriculture because it is associated with addressing climate change, which many don't believe in or see as a liberal lie meant to push them out of agriculture. And while I trust this percentage of farmers is small, some do not think long-term and instead adopt a "take all you can, while you can" attitude toward our shared soil resource. Transitioning to regenerative agriculture also involves a stated or unstated admission that industrial agriculture isn't the best option for producing food in a changing climate—and we all know how hard it is to admit we're wrong, whether to others or just to ourselves.

What most farmers *do* accept is the corn-and-beans mentality, a shorthand reference to grain production relying on tillage or chemical no-till, fertilizers, GMO seeds, cash crop rotations without covers, and other industrial strategies. Why the corn-and-beans mentality reigns is no secret: Big Ag propaganda combined with government programs incentivizing large-scale commodity production convinced farmers they'll go broke without conventional inputs, a tight focus on yields, and specialization. As rural anthropologist Jane W. Gibson and her colleague Benjamin J. Gray describe it, "Farmers have been ideologically colonized by the values of industrial agriculture."[8]

The corn-and-beans mentality is one reason many Midwest farmers react to regenerative practices like the nosy neighbor. Another is the simple fact that Susan, Dawn, Cathy, Kelly, and Sariah are women, and sexist attitudes linger about a woman's place in agriculture. Researchers have documented the social and institutional biases against female operators, especially as related to conservation.[9] Sometimes these biases arise from the perception that the tasks women engage in do not constitute farming. Growing up in rural western South Dakota, I watched women like my mother work on farms and ranches. These women labored just as hard as their husbands. They put in a full day moving cattle or stacking bales, and then made dinner and cleaned the house. They served as accountant, gardener, cook, errand runner, and barnyard livestock tender. They handled basically everything related to the kids. Although their work held the operation together, the dismissive term for these women was "farm wife."

In the white Western agricultural tradition, women have long done farmwork, but men typically claimed the title of farmer or rancher because men did most of the "real" physical work. To this day in many conventional agriculture circles, women are usually understood as helpers or hobby farmers, not as farmers or agriculture leaders in their own right.[10] "Women have been invisible in this space for a long time, but they've been very much a part of the ag world," Roesch-McNally says. "I still laugh when I talk to women who are like, 'Oh, I'm not the farmer, that's my husband, but I've been doing the farm books.' In any other business if you did all of the accounting and finances, you would consider yourself a critical component of the business. But in agriculture, it's different."

The kind of thinking Roesch-McNally describes goes all the way back to the invention of the plow. For thousands of years and across the globe, writes Mark Bittman in *Animal, Vegetable, Junk: A History of Food, from Sustainable to Suicidal*, women were the primary agriculturalists: planting, seed saving, domesticating animals, developing tools, and determining how to use the land. Men and women divided labor, but "these divisions were not defined by dominance and subservience . . . it seems likely that roles shifted toward patriarchy as the plow and other heavier equipment that required significant brawn were introduced to farming."[11] In other words, when farming became more about physical than intellectual work, it also became more male.

Patriarchal institutions went on to support this development in America. From colonization forward, farms typically belonged to men both on paper and in the minds of their peers. Land often passed from father to son and bypassed daughters. Partly to blame is the rural idyll myth, which casts rural life as romantic and centers the heterosexual white couple and their children as natural farmers. "Under this vision, (white) men are the farmers, and their (white) wives are their helpmates. When the time comes, the farmer's son inherits the land, and their daughter either disappears from the imagination or is expected to marry a (male) farmer or rancher if she wants to stay in agriculture," write scholars Ryanne Pilgeram, Katherine Dentzman, and Paul Lewin. They go on to argue that "part of the work of rewriting women into agriculture is about dismantling this fiction and yet recognizing that it creates real and persistent barriers for anyone who exists somehow outside of the rural idyll trope."[12]

The rural idyll is not only a sexist and racist fiction, but also an inaccurate depiction of the economic reality on most farms. Countless female farmers work full- or part-time nonfarm jobs to keep family finances afloat and secure health insurance. They squeeze farm work into whatever time is left. Many of my farm friends' moms lived this kind of life back in South Dakota. Willingly or unwillingly, these women traded on-farm hours for steady paychecks that covered seed, fuel, and livestock feed, often because farming conventionally as their husbands were determined to do is financially tenuous. They ended up in a double-bind: the women's off-farm work further robbed them of the title of farmer or rancher, but they could not fully dedicate themselves to their careers (and thus earn complete legitimacy in the workplace) because they shared their energy with the farm. Some women bore the burden of the rural idyll more than others over the years. Widely documented systemic racism often pushed farmers of color, particularly Black farmers, out of agriculture or prevented them from entering altogether.[13] Unmarried or queer female producers were rare, and those who were on the land often faced discrimination and outright hostility.

These combined social norms—the "farm wife" label, the unseen labor, the rejection for embracing sustainability or otherwise not fitting the white, heterosexual mold—remain largely unchanged in industrial agriculture, a sector still controlled overwhelmingly by white men. That they dominate the agribusiness world should be no surprise. After all, white men implemented settler colonialism rooted in slavery and the removal of Indigenous tribes, then built the postcolonial American agricultural system that became modern conventional production. They adapted European philosophies of dominion and extraction to the landscape, then applied American industrial and capitalistic practices—economies of scale, mechanization, standardization, genetic engineering, chemical solutions. For most of this country's history, men controlled federal and state agriculture programs and policies, universities and private labs devoted to agricultural science, countless businesses related to food, and land itself and decision-making power over it. The result: industrial agriculture and the infrastructure to support it stretching from coast to coast.

Agriculture in this industrial form came to be about size, another distinctly male obsession. Bigger acreages, machines, yields, and herds. Bigger loans to finance these must-haves. Agriculture also harnessed the familiar patriarchal concept of control: of weeds, insects, and soil through chemicals, of farmworkers through low wages and deportation threats, of larger producers over smaller ones through buyouts, of uncontrollable weather and markets through industrialization. Capitalism drove the entire transformation—the profit seeking, the exploitation, the zero-sum competition between farmers and their environments and between one another as producers.

Farms started to look more like factories. Industrial agriculture features "production practices that . . . tend to emulate modern factories: industrial farmers specialize in the commodities they sell; they operate in a highly competitive, global market; they rely on sophisticated machine, chemical, and genetic technologies; and they pursue efficiency and profit, necessitated by a global capitalist system."[14] Industrial thinking is how we ended up with livestock finished in concentrated animal feeding operations, extreme land consolidation, hollowed-out rural communities, agrochemical dependence, and widespread ecological devastation on and near agricultural lands.

Breaking the conventional, capital-driven mindset—seeing through its lies and envisioning something holistic, ecological, nutritious, inclusive, and socially just instead—is perhaps the largest barrier to building a regenerative food system. But over and over, people within the movement tell me women are uniquely talented at removing the Big Ag blinders. Growing female participation in regenerative agriculture might just create the tipping point needed to pull back the curtain on conventional agriculture and remake the system. "I look at women to be the saving grace on this, because women are more likely to carry out and continue conservation practices, whereas the men tend to look at the dollar signs," Susan tells me in an earlier conversation. "The women are looking at, how's that going to be next year if we do it this year? I think that's why I see getting more women involved will actually change the scope and progression of regenerative ag."

Just then, Kelly's cell phone rings; it's her husband calling to warn her about possible tornadoes in the storm, whose clouds suddenly look quite close and dark. "The weather radios are going off," she tells us. The wind picks up as we hustle back to Dawn's house. We chat in the garage until gusts drive the rain in sideways through the open door. Shortly after we retreat to the living room, hailstones drop from the sky and lightning flashes too close to the windows for comfort (mine at least). I'm picturing my rental car back at the truck stop, festooned with dents or a broken windshield. The power goes out and does not come back.

When the storm finally passes, Susan and I strike off in her truck. We encounter flooded roads, crews fixing power lines, and downed trees, but no signs of a tornado. Susan stops at a badly eroded field, where water has carved deep scars in the hillside and pushed topsoil into a pile near the road. A pool of muddy water sits stubbornly at the hill's bottom. The field has nothing growing on it; presumably it is "resting" as summer fallow, kept free of all plant life for a season thanks to herbicides. Summer fallow is a practice that provides the opposite of rest; instead, it stresses the soil because soil needs living roots to stay alive and in place. The erosion I see is exactly what a cover crop can help prevent, Susan points out. "Even if it would have been three inches tall, that would have been three inches of roots underneath, and it wouldn't have washed near that bad," she says. "Look at that puddle. That's not a puddle; that's probably four inches deep."

We make it back to the truck stop and say our goodbyes. I'll head to Kansas City for the night, then to Wichita in the morning, then to planes that will carry me back to my adopted home of South Florida. I think about the bigger journeys we're on as a nation, journeys both exciting and dangerous. The one we're taking away from conventional agriculture and toward regenerative. Away from the "normal" world we've known and into the realities of a warmer planet. For now, though, I aim the car west, into the dazzling glare of a June sunset in farm country, the possibilities these fields contain stretching all around me.

To gain a deeper sense of Susan's impact, I return to Missouri two years later to see an example of how she helped a conventional row crop operation go regenerative. On a mild mid-April morning she and I convene at Rusted Plowshare, a three-hundred-acre farm outside of Columbia owned and operated by Josh Payne, forty, and Jordan Welch, thirty-three. This brother-sister duo converted their family's corn and soybean farm into a regenerative organic lamb and beef business—a tremendous feat given the farm's history, the Payne family dynamics, and the grip of industrial agriculture in the Corn Belt.

Like many midwestern producers, Josh and Jordan come from generations of farmers. Great-grandparents and grandparents on both sides farmed. Their maternal grandparents, Charlie and Hazel, bought in the 1950s the acres they currently operate. Their parents, Jon and Debra, raised them on the land, where Jon worked part-time alongside Charlie and full-time in banking—until Jon tragically passed away in a farm accident in 2002, when Josh was nineteen and Jordan twelve. It wasn't the family's first loss; Jon's sister, his only sibling, had passed away a little more than ten years prior. "That middle generation is just completely gone," Josh says about the paternal side of his family.

With their link to agriculture all but severed, Josh and Jordan left Concordia after high school to pursue careers in education. Josh and his wife, Larin, lived in Kansas City for a while, then moved back to Concordia in 2010 to teach school and raise their children. A few miles away, Charlie was in his eighties and farming alone. Josh decided to help his grandfather part-time, and later quit teaching to farm full-time—but he quickly soured on the work.

"It was literally something I despised, like I hated every minute," Josh says. "It was just your conventional corn and soy. It was lots of tillage, lots of chemicals. I didn't know anything. I wasn't good at it. It's not working with people at all, it is working with tractors and equipment and having to know what chemicals you're putting on and what seed, all the stuff that wasn't terribly interesting with farming."

Josh practiced agriculture according to conventional rules, rules Charlie had embraced long ago and adhered to fiercely. Josh yearned for something more. "At some point in there, I found out about cover crops," he says. "I had three years of arguments with my grandfather about cover crops. And then we finally tried it, and they worked really well. That created this sense of meaning. In that process, I realized that farming could really be meaningful because I saw this concept that, it's not just about making money. The phrase that I use when I go to talk places often is, commodity farming is just growing nickels and dimes. You're not really grow-ing food. So I started to think about how I can restore soil instead of just killing everything." Adding cover crops to the corn and soybean rotation was progress, "but it still didn't really fit where we wanted to be long-term," Josh says. The farm remained a row crop operation at heart with little diversification, and Charlie balked at incorporating further regenerative practices. But a breaking point was on the horizon.

For years, Josh had been experiencing what he believed were allergic reactions to food. His throat would swell shut with phlegm, so much that he could barely breathe. A doctor finally asked Josh to look for patterns in these reactions, and he realized he'd endured a spike of incidents during high school that receded when he left Concordia but returned immediately upon moving back. The reactions grew more numerous and intense; during harvest season, for instance, he suffered up to three reactions a week. In February 2020, the doctor diagnosed Josh as allergic to agrochemicals, especially to the glyphosate in Roundup. His body was responding with quasi-anaphylactic shock when he spent significant time in the fields, and the doctor predicted these reactions would eventually kill him. Josh and his grandfather faced a choice: either change the way they farm, or Josh needed to leave the operation.

Josh knew he had to present Charlie with a solid plan. Sheep came to mind, since a few years earlier with Susan's help he had experimented with raising a handful of sheep and he knew the economics were promising. He asked Susan for assistance with grazing plans and logistics, and he and Jordan continue to seek her advice on animal and soil health issues today. "I was like, 'Susan, what do I do?'" Josh recalls. "Susan was really important in, how do we put this stuff together? Pretty much without her, one, we wouldn't have had the idea to do sheep. But two, they all probably would be scattered all over the country somewhere, so it wouldn't have worked out." The returns on sheep and the prospect of losing Josh convinced Charlie that regenerative organic livestock production could work.

By March 2020, two hundred pregnant ewes called the farm home and Josh was racing to build fence. For help he turned to Jordan, who lived and taught school in Lawrence, Kansas, at the time. Jordan had zero agricultural experience. Charlie had prioritized Josh's presence over hers, assuming Josh would buy her half of the farm eventually. But Josh recognized the talents and perspectives Jordan could offer. "Jordan is so much better than me at so many different things," he says. "By having more than one person on a farm, it completely changes what we do and how we do things. Without that other gender, without those other voices, I don't think we would be as successful as we have been."

Jordan agreed to build fence over the summer. She loved the farm so much that she and her husband soon moved to Concordia, where she secured a new teaching gig and continued working part-time on the land. "I had no intention ever of moving back to Concordia, let alone the farm," Jordan says. "But I just found I really enjoyed being outside and I really enjoyed the manual labor of it. It was peaceful and relaxing, and I know in teaching there's always a lot of stress, and especially now. [The farm] seemed like it fit my personality better, and I liked the way that the farm was moving." By that she means the transition to regenerative organic. Jordan was already interested in ecology, nutrition, and environmental stewardship thanks to her role as a mother. "All the stuff that you do affects [kids] and their health," she says. "My first child has asthma and allergies, and a lot of that can come back to the stuff that we eat and the stuff that we put on our skin. So I think that's kind of what started me in that direction. Kids change everything."

That connection Jordan sees between food and health is also apparent in Josh's agrochemical allergy. But although his reactions were painful, perhaps they were a blessing in disguise, since they forced a full conversion to regenerative agriculture that may not have happened until after Charlie's passing. Many farms have a Charlie: a family elder entrenched in industrial agriculture who maintains a tight grasp over farm decisions. Men like Charlie are an outgrowth of the patriarchy baked into conventional agriculture. They might mean well by opposing change that, in their view, threatens the farm, but too often their recalcitrance either perpetuates the industrial system or drives away younger generations, especially their daughters and granddaughters. Charlie wasn't happy when Jordan joined the operation and has always seen Josh, not her, as the farm's natural manager. "It wasn't until he saw me building all that fence that he realized that I could do this. I'm not as fast as

Joshua is, but I can keep up with him, and Grandpa didn't believe that at first," Jordan says. "Grandpa still probably doesn't fully support me working on the farm."

Charlies of the world aside, the conventional model itself discourages young people from staying on the farm because, in many cases, there is simply no financial room for multiple families. According to the latest Census of Agriculture, average net annual farm income was $43,053, down 2 percent from five years before. Farm income includes sales, government payments, and earnings from farm-related activities.[15] Imagine how low that income level could drop if government aid wasn't factored in. The same year of the census, 2017, average annual per-farm expenditures were $176,352, a 4.3 percent bump from the prior year. For crop farms, that expense average was even higher at $210,081.[16] Most conventional commodity producers require operating loans and other forms of credit and government support to stay in business. Going regenerative means losing some or all of that government aid, as well as access to government and private financing.

Most of our farmers don't have the money to front a regenerative transition, even though such a change would generate more income per acre as soils recovered and input expenses dropped, which would in turn create financial room for the next generation to stay on the farm. Data proves the economic rationale behind regenerative agriculture. Researchers who evaluated regenerative corn fields in South Dakota, North Dakota, Nebraska, and Minnesota recorded 78 percent higher profit per acre compared to conventional corn fields.[17] Still, the psychological transition from industrial to regenerative is a barrier, as is the lack of institutional support for such a change. "It's more of a mind shift," says Jessica Hulse Dillon, director of the Soil & Climate Alliance. "It's getting the farmers to understand that, yeah, there might be a year or two where you could see this drop in yield. But the benefits afterwards are exponential. However, when you're talking about farmers who are every year riding on the edge of losing their farm, a year or two could really break them." If we fail to provide farmers with financial resources to transition, then we will miss an opportunity to bring young people back to the land, to farms profitable enough to sustain them and their families.

In addition, many of the farm communities young people would remain in have stagnated thanks to population loss driven by farm consolidation—schools closed, businesses departed, social life diminished. Staying looks unappealing from an environmental perspective as well; it means working on dead soil, enduring ever more erratic weather events, and using conventional tools like chemicals and GMOs that exacerbate both problems. It could mean living near concentrated animal feeding operations or drinking water contaminated by agrochemicals. "The whole thing of how we have depleted soils, if you look at it, how many families are depleted?" Susan says. "The next generation is depleted simply because it's so hard and mentally devastating to work on a farm that has no resilience. That next generation is like, 'There's no future here. Why would I stay here?' And the parents are unhappy and angry, but they're stuck in this or they think they are. It is just monetarily impossible to think of having extra people that need to be paid on the farm, and that generation sees that."

Rusted Plowshare is the opposite of what Susan describes. Josh and Jordan expanded the farm's enterprises to include direct marketing of lamb and beef products, chestnut production, and occasional farm-to-table dinners. Susan has been a key ally in their efforts to scale up and diversify. The farm now generates enough income for both Josh and Jordan to work there full-time without other jobs, a move their spouses might make someday. "We can hopefully on these three hundred acres have six full-time jobs," Josh says. "If we do that, then we've gone the other way of Big Ag. It's gone up; now it takes about two thousand acres to be a row crop farmer in our area. So that's one job on two thousand acres. But if we can have six on three hundred acres by doing intensive grazing and chestnuts and other things like that, then we've added five jobs to the local economy." Jobs mean families—families that attend schools, shop at stores, eat at restaurants, buy homes, pay taxes. Families with a connection to their food, with children who learn how to raise animals and grow gardens and maintain soil health. "Josh can bring his kids out here, Jordan can bring her kids out here, and let them see what's going on," Susan says. "They will be the next generation that understands—not just learns but understands—soil health and livestock."

My tour of Rusted Plowshare begins as such visits usually do, with one exception: a film crew is producing a documentary about Josh and Jordan's regenerative agriculture journey, and they're shooting this morning's adventures. The sound recordist mikes the four of us up. The director explains that she and her fellow film-maker are utilizing cinema verité, or truthful cinema, an obser-vational, fly-on-the-wall style where people are supposed to act as if the camera is not there. I find this mandate difficult; it's hard not to feel like I'm performing, even harder to forget the crew's presence. What's more, I am in the strange position of interview-ing people for a project while simultaneously being a subject in someone else's project. Suddenly I have a deeper understanding of what interviewees might feel when talking to me, a more profound gratefulness for their trust. Sharing your story comes with the risk that its retelling will be flawed, or even mean-spirited. I trust these filmmakers; our projects have similar goals. But I don't trust myself not to do or say something embarrassing on camera.

Within minutes my hope of a humiliation-free morning evaporates. Josh, Jordan, their dog Roxie, Susan, and I climb into a side-by-side and rumble out to the field behind Josh's house, where they're keeping the beef cattle at the moment. Like Kelsey, Josh and Jordan rotate their herds across the land, but their rotations are daily. Another difference: their herds graze fields planted with cover crops rather than prairie pastures, although Josh told me earlier that they are converting several fields back to native grassland. Josh and Jordan moved the sheep before I arrived, but the cattle need fresh forage, so I gamely agree to help erect a new line of portable electric fence.

But first we need to feed the cattle their farm-grown microgreens, which will distract them while we set up the new fence line. Jordan handles the microgreen fodder operation. The process is basically allowing long trays of barley and cereal rye seeds to sprout, and letting those sprouts grow until they are big enough for livestock to eat. She sources the grain from local farmers, a boost to the area's economy, but as Josh says, "The plan is to eventually make it so that we grow our own stuff, close the circle." Earlier this morning I watched Jordan spread grains evenly on trays and arrange them on shelves that stretched ceiling-high and wall-wide. Drip lines watered trays in various states of growth. When the sprouts reach six inches high with four-inch roots, usually within six days, they are ready for consumption. Microgreens are five times more efficient than grain at providing the energy livestock need to put on weight that becomes well-marbled, tender, rich-tasting meat. The animals also maintain their status as grass-fed because greens are classified as grass under the rules of USDA organic certification. Organic rules allow ranchers to feed organic grain in an animal's final months, but to Josh and Jordan feeding grain dilutes the authenticity of grass-fed. The greens don't replace grass and cover crop forage in the growing season or hay in winter, but they do serve as a daily year-round supplement.

Moreover, the practice illustrates how farmers and ranchers can adapt regenerative principles to meet their needs instead of molding their operations to fit a status quo. Microgreens are an innovative way for their cattle and sheep to eat living forage right up until slaughter while keeping their cost per pound as low as possible. Plus, Rusted Plowshare has buildings and available labor for growing greens, and the herd is the right size for such a hands-on fodder program, so the practice is a natural fit. The daily chores involved also work in tandem with the daily livestock rotations. And while feeding microgreens may sound strange to conventional ranchers, their production and distribution parallel typical fodders, like hay or silage. How Jordan grows and harvests livestock feed looks different—she sows and reaps daily and on a smaller, indoor scale, as opposed to planting hay crops, driving a swather, and making bales—but the essential tasks are the same. While we ride, Jordan dumps out mats of microgreens from the manure-spreader-turned-fodder-distributor that's hitched behind the side-by-side. The steers gallop and buck in anticipation, then quickly settle in to eat.

Now the clock is ticking to construct that fence. Jordan demonstrates how to set the posts and unspool the wire, then hands me the reel. I grip the handle and walk forward awkwardly, the four strands of wire unrolling behind me. Tangles ensue that Jordan must unravel. Whenever I press my foot down on the fenceposts to drive them into the ground, the slick soles of my cowboy boots slip off the plastic foot bars. The wind blows my too-short-for-a-ponytail hair into my eyes. The whole process takes me triple the time it would have taken Jordan, and the fence ends up wildly crooked. Then she shows me how to take down the fence we don't need anymore. I should point out that I'm neither strong nor coordinated. My holding the reel in one hand and with the other turning the crank that spools up the wires, while walking forward and keeping the four strands taut so they don't tangle (which they do, over and over), will prove comical on-screen, if the moment makes the cut.

With the steers rotated, we ride to see Josh and Jordan's flock of three hundred ewes and their lambs. A sheepdog trots over to greet us, and lambs bleat as we walk through the herd. The land we are on, and much of what we call the Corn Belt, was once tallgrass prairie adapted to herbivore and fire impact. According to the Missouri Prairie Foundation, prairie covered about one-third, or at least 15 million acres, of present-day Missouri until its statehood in 1821. All that remains is 51,000 scattered acres, about half unprotected. What Brazil is on track to do in the Amazon rain forest is what America already did to the tallgrass prairie: entirely convert an expansive, carbon-sequestering, weather-making ecosystem rich with animals, insects, plants, and Indigenous tribes into a carbon-emitting monoculture managed with agrochemicals and diminished of ecological life.

But as I am seeing, rotational grazing and cover crops help the land recover from such farming and re-create the effect of bison, fire, and Indigenous stewards. This is possible even in the Corn Belt, the most intensely and chemically farmed region of the United States. Soil carbon levels average around 4.5 percent in Rusted Plowshare's fields, up from measurements eleven years ago of about 2 percent, a figure that remains the average for the area's soil type today.[18] Insects, birds, and other wildlife are returning. Rotating the sheep and cattle disrupts the parasite cycle, which eliminates the need to deworm with parasiticides. Holistic grazing also increases the nutritional quality and diversity of the forage so that the animals resist illness more effectively. "They get to go out and pick what they need," Jordan says. "There's lots of diversity in this field, so they get to pick what nutrient-wise is good for them."

These animals are healthy and content, a far cry from what I would see in feedlots where most sheep are finished—feedlots propped up by the cheap industrial corn and soybeans Josh used to grow. Only yield matters in that system, not nutritional quality, since corn and soy are destined for animals, ethanol plants, or Big Food processors rather than direct human consumption. It's all about bushels per acre. "When I was growing corn and soy, I was really just growing nickels," Josh says. "I wanted it to be food, but it wasn't really food because it was going to an ethanol plant. It doesn't really matter what you put into it if it's going into something else, whereas these guys"—he gestures toward the sheep—"I've got to really care for them because I'm going to sell them to somebody to eat. I'm not just growing nickels anymore, I'm growing animals. I think that kind of changes a person's perspective."

Next stop is the thirty-acre chestnut orchard. Most of the trees are about six years old, and the first harvest yielded roughly five hundred pounds of chestnuts. By about year thirteen the orchard should generate around $200,000 in profit a year, the equivalent of about a thousand acres of corn and soy with far fewer inputs, labor, and land. An orchard is a sight I have never seen on a Midwest farm, but not because the region is unsuited to fruit and nut production. Rusted Plowshare's latitude is ideal for these grafted Chinese chestnuts that bear every year and demonstrate exceptional resiliency. Rather, I occupy the wrong era. Orchards went the way of fencerows, livestock, gardens, and crop rotations with the rise of industrial agriculture. In 1920, roughly one hundred years before today's tour, "84 percent of Iowa farms grew apples, 62 percent grew potatoes, more than half grew potatoes and cherries, and nearly a third grew plums and grapes. Other relatively common crops included strawberries, pears, and peaches. By 1997, not a single fruit or vegetable crop was grown on more than 1 percent of farms."[19]

Many Missouri farms once included orchards too. The historical society in Cooper County, just west of Concordia, documents that German immigrants planted orchards when they arrived in the 1850s, and that Charles Bell of Bell Apple Orchards founded the International Apple Shippers Association in the county.[20] When grandpa Charlie purchased the farm sixty years ago, an apple orchard stood in the exact spot where the siblings planted their chestnuts. The orchard's placement was pure coincidence. No one knew about the orchard demolished long ago to make way for corn and soy until Charlie teared up as the chestnut trees were going in.

Josh and Jordan's version of regenerative is just one option for Corn Belt and other growers. Farmers could keep growing cash crops like corn, wheat, and soy, but incorporate cover crops before and after. They could fold additional cash crops like barley, canola, or oats into the rotation. They might use polycropping, also known as intercropping, to grow two or more cash crops together. Or companion cropping, where a cover crop grows along with a cash crop. Or strip cropping, which means planting alternating crops in long strips. They could devote some land to perennials like elderberries, Kernza, or alfalfa, switch to no-till planting, add riparian buffers, introduce bees, and use compost as fertilizer. They could bring in different moneymaking livestock, such as goats, chickens, pigs, geese, or dairy cattle. Perhaps the most exciting (and lucrative) option is converting Great Plains cropland back to native grassland and selling grass-based products like native plant seed, hay, and livestock--a successful economic and environmental model documented by the EcoSun Prairie Farm experiment.[21]

One size truly fits none with regenerative, so long as farmers follow basic principles: minimize soil disturbance, keep living roots in the ground, cover the soil, increase biodiversity, use animal or natural fertilizer, and work within the context of the surrounding environment. I would add social regeneration, too, in the form of fair wages for farmworkers, equal opportunities for all producers, recognition of and remedies for racial and gender injustice, ecological restoration, and pathways for young people to enter agriculture.

I leave Rusted Plowshare feeling optimistic. Here is a family that once epitomized conventional row crop production, and now proves the economic, social, and human health promise of regenerative agriculture. If the Paynes can make the regenerative transition in an industrial agriculture stronghold, and welcome back the lost female generation of farmers, then almost any family can. Having an institutional partner like Susan made a crucial difference in their success. Imagine how the regenerative movement might progress if more people like Susan existed within the government, university, and private entities that interact with farmers and ranchers. Imagine the direction our nation might take if we followed the example of the Payne family and others like them in embracing change, chasing abundance, and prioritizing the next generation.

4

Motion

Bu Nygrens, Mary Jane Evans, and Karen Salinger, Veritable Vegetable

Motion: the traveling pattern exhibited by a wave.

My younger brother Joshua Johnson raises beef cattle in western South Dakota. His goal: use regenerative grazing to restore the native prairie ecosystem on 1,200 acres of formerly conventional pastureland. He also employs cover crops and rotational grazing to build soil health on 160 acres of formerly industrial cropland.

Josh sells as many beeves as possible direct to consumer but can't market his entire calf crop that way. Right now he sells about fifteen animals per year in wholes, halves, and quarters based on consumer demand and available slots at local slaughterhouses. He typically increases his number of direct-to-consumer animals by one each year. The rest of his annual calf crop ends up at the sale barn, where the auctioneer informs the crowd that the calves are antibiotic- and hormone-free, entirely grass-fed on pesticide-free land, and never confined. But the buyers, unable to offer a premium, pay the going rate for live feeder cattle and the animals end up at a feedlot. The consumer will never know the difference. A similar situation occurs with grain in years Josh decides to incorporate cash crops into his rotation—there's nowhere to sell it except area grain elevators, where it receives no premium and is dumped in with conventional grain.

What a shame, I thought as Josh and I herded his cattle from a cover crop field to a fresh pasture when I visited him and my parents the weekend before my tour of DX Ranch. The day's temperature had pushed 90 degrees, so we waited until the cool of evening to move the herd, Josh on his dirt bike so I could drive his ATV. The late September sunlight was soft and forgiving, the air dry and dusty from drought. Consumers won't know that Josh's cattle are fleshier than others I've seen in the area on this trip, thanks to the relatively abundant forage in his pastures. They won't understand the carbon these animals help the soil sequester, or the high nutrient density in the meat they will become. They won't see the human, my brother, ranching regeneratively on principle, without the market and institutional support conventional producers enjoy. Watching Josh's cattle stream through the gate and into the pasture, I am intensely proud of him. By going regenerative, he's positioning his business and land to endure climate change more effectively.

But there's one problem Josh can't solve, a problem that keeps many farmers and ranchers from transitioning to regenerative agriculture: the lack of a robust regional food system to absorb and pay a premium for large quantities of regeneratively grown grain, meat, and other products, keep them separate from their conventional counterparts, and identify buyers who will pay more for those outputs. Unlike organic, regenerative has no USDA label, although third parties like the Rodale Institute and A Greener World have developed regenerative certifications. In the grocery store, an apple grown using climate-friendly regenerative practices looks exactly the same as one grown with agrochemicals, synthetic fertilizers, and other industrial methods.

Some farmers can opt out of the conventional supply chain. Carrie and Erin are examples, since they sell all of what they grow through a CSA, food hubs, and farmers' markets. Josh and Jordan, as well as Kelsey, market a portion of their output directly to consumers, too. Farmers and ranchers engaging in alternative marketing like this do an excellent job of feeding communities and bolstering local economies. They're important for a thriving food system. But not everyone can sell directly to consumers or other nontraditional markets. Some producers live too far away from cities and towns, or simply are not interested in that approach. Direct marketing requires an entirely different skill set than farming and ranching, not to mention a good deal of time to do it right.

Plus, most of America's farmers and ranchers produce too much of whatever they grow to sell it all directly to consumers—their operations are just too big. For commodity agriculture in particular, what farmers harvest—such as unprocessed sunflowers, canola, wheat, corn, or barley—usually isn't suitable for selling right to consumers in the first place. Transforming outputs into market-ready ingredients or products is too complicated and expensive for individual growers; most can't, for example, build their own mill, grind the wheat, and sell the flour. For many, such an endeavor is outright impossible given their location and the lack of nearby food system partners to help.

Also, many regenerative farmers are forced to keep growing the same cash crops as conventional farmers, just with different practices, again because the food system isn't set up to absorb other crops as readily. Industrial food and farming are hyperfocused on growing and processing a handful of grains: corn, soybeans, and wheat. Alternative crops that would bring diversity to the land, a key regenerative principle, are almost impossible to market in that system. Jessica Hulse Dillon, director of the Soil & Climate

Alliance, ran into this very issue seven years ago. She and other partners were exploring what a supply chain built around small grains might look like in the Midwest. "What we learned in that process is, it could be done. It was actually not that big of a lift to get farmers to grow small grains, it wasn't even necessarily that big of a lift to get some end users to use the small grains," Hulse Dillon says. "Where the entire thing fell apart was the middle of the supply chain. We couldn't move them, we couldn't store them, we couldn't process them. Those facilities did not exist."

In a perfect world, all food would be regenerative and we wouldn't need labels or separate supply chains. "What we have to do to move the supply chain is we need to get to a mass balance. So we need there to be so much regenerative wheat coming into a processing facility that it's regenerative wheat. If a little wheat gets in there that isn't, it's fine," Hulse Dillon says. "We are really looking at how do we get so many people doing this that we're not necessarily having to treat this like an organic or a non-GMO, something that has really strict testing standards on it. Because that's not how we shift the system." I truly hope what Hulse Dillon describes comes to pass; she and many others are working hard to make it so. But we are a long way from that world of mass balance.

Right now, what my brother and regenerative producers like him need are like-minded supply chain partners—distributors, processors, food labels, cooperatives, or other entities in their regions—to purchase and transform what they grow, in Josh's case live cattle. Ideally that partner specializes in regenerative products and builds lasting relationships with growers. Verification might be part of the process, but in exchange farmers and ranchers could receive price premiums they otherwise wouldn't in the conventional market. And producers would know exactly where their outputs go within the food system and that the regenerative integrity remains.

I wanted to see what such a partner might look like. How would they source, create, and deliver products? How might they enact their values? Was their business economically feasible? My quest took me far from my brother's South Dakota operation in both place and time—all the way back to 1970s San Francisco, in fact.

The year is 1978. New York City native Bu Nygrens is twenty-seven years old and has lived in San Francisco for about four years, working odd jobs as twentysomethings do, but with an ecological bent. This was the era of Frances Moore Lappé's *Diet for a Small Planet* and Fannie Lou Hamer's Freedom Farm Cooperative. Americans were growing concerned about agricultural pollution, the connection between poor food quality and disease, and notions of what we today call food sovereignty and food justice.[1] "Coming from a big city, I was personally sort of a 'back to the lander' type and I thought my destiny was going to be on a hippie collective farm someplace on the West Coast," Bu explains. A friend of Bu's worked at a produce distributor called Veritable Vegetable, which operated with the grounding belief of "Food for People, Not for Profit." The company supplied produce for the People's Food System, a collective network that brought affordable, nutritious food to San Francisco food stores. Bu's friend offered her a gig driving a truck a few days a week. "I said, 'I've never driven a truck. That sounds fun. And besides, you work with farms. Maybe I'll find out where all the cool farms are,'" Bu recalls saying to her friend.

By 1979, Bu was working full-time for Veritable Vegetable. Two other women, Mary Jane Evans and Karen Salinger, were also on the VV team. VV had a trucking operation and its own warehouse and was supplying produce for dozens of retailers. "Organic" was not yet an official USDA label, but the company prioritized pesticide-free, sustainably grown food anyway. Women were increasingly finding a professional home at VV, drawn by the likes of Bu, Mary Jane, and Karen and the welcoming environment they helped create. Paying farmers fair prices; offering equitable wages, benefits, and equipment for staff; and contributing to environmental and social justice were the company's priorities. Through the eighties and into the nineties, VV amplified its connections with farmers, increased its emphasis on organic growers, expanded its trucking division, and offered commercial freight service to avoid empty backhauls and fuel waste.

Bu relays this history to me over a bowl of fresh cherries in VV's current offices in San Francisco's Dogpatch district. The story is more than simple background information for me, a curious writer. "Food for People, Not for Profit" is still a guiding principle. The city's famed progressive social movements shaped Bu, Mary Jane, and Karen in profound ways, helping them turn 1970s ideals into lasting realities. They developed values related to food, feminism, the environment, and social justice, and they brought these values into their leadership of a company that models what a regenerative food system might look like.

When I began looking into such a system, produce distribution seemed dizzyingly complex—and it is thanks to unpredictable variables like weather, crop performance, prices, and labor availability—but Bu provides a straightforward overview. "We buy from farmers," she said, "and then we bring the product back here and then we sell it in smaller quantities. So we buy fifty cases to three hundred cases of one thing at a time, depending on how the pallet is built and how big the boxes are. We then sell in single-case units to our customers. For smaller customers, we may break the cases." VV has a diverse mix of customers, the largest sector being retail. Others include restaurants, caterers, food service operators, manufacturers, juicers, corporate campuses, box or meal kit pro-

viders, schools, hospitals, wholesalers, and freight customers. VV
features midsize farms on its product list but supports both large
and small producers. Today VV delivers organic produce and other
select organic perishable, floral, and grocery items all over California and to parts of Arizona, Colorado, Nevada, and New Mexico.
They also ship to Hawaii and New York. Working with hundreds
of growers, some direct from farms and some through packers
or shippers, they offer more than 750 different source-identified
organic fruits and vegetables daily.

The diverse product list and grower network are strengths that
set VV apart from bigger competitors, Bu says. "Because we work
with small growers in particular, it would be much more efficient,
effective, cheaper, if we did like other mainline distributors do,
which is you offer one size, two items, and those are the choices.
So when somebody buys an apple they don't get to select between
several. Our price list, what we offer, is seventeen pages long every
day because we give customers the choice between not only different varieties in different sizes, but different farm sources. Because
we have source-identified product, we might have kale from six
or seven different farms at the same time. A broadline distributor
like Sysco or US Foods or somebody like that, you have no choice
between this kale or that kale, maybe organic kale or nonorganic
kale if the price is different."

Distributors like VV are not the norm in America's food system.
Instead, national and global corporations circulate most of what we
eat, companies like Sysco and US Foods, which Bu mentioned, but
also others like Performance Foodservice, Gordon Food Service,
and McLane Company.[2] When it comes to produce, meat, dairy,
and other farm goods, these companies work almost exclusively
with big, conventional growers that supply large volumes of con-
sistent products. Because these companies set the prices—and set

them low in many cases—farmers stay financially afloat by operating industrially: monoculture cropping, pesticides, synthetic fertilizers, continuous cropping and grazing, rock-bottom wages and few benefits for farmworkers, heavy machinery use, feedlots, cheap livestock feed, and so forth. For most distributors, profit is king, gained at the expense of the environment, farmers and their workers, and the food's nutritional quality.

In contrast, VV is a regional food hub, according to the USDA's definition. A regional food hub is a business or organization that actively manages the aggregation, distribution, and marketing of source-identified food products primarily from local and regional producers to strengthen their ability to satisfy wholesale, retail, and institutional demand.[3] "Food hub" is a relatively new term to describe what VV has done since its inception. Food hubs take a triple-bottom-line approach, striving for economic, social, and environmental gains simultaneously, particularly regional gains. "We created that model without calling it that until after the fact," Bu says. "The food hub model as purported or promoted by the USDA and other world development agencies and ag economists and people like that has always been that regional control of distribution is more appropriate. Short value chains are agile and impact local economies more than long supply chains, where food is considered a commodity and prices are forced down. Leveraging your resources and consolidation definitely does make things cheaper, but does it make them better? Are diversity and quality and value more important than price and quantity?"

Compensating farmers fairly, encouraging them in sustainable practices, and adapting to the farm's needs rather than dictating policies are a few ways VV and similar distributors support regenerative production and rural communities. Growers are partners, not just exchangeable cogs in a profit machine. Like the cherries we're snacking on: Bu tells me exactly what farm they came from, describing family members there she's known for decades. She relays how climate change and labor shortages have collided to disrupt their business. As weather patterns vary and temperatures warm, the farm's cherries and peaches now ripen at the same time. Worker shortages, worsened by restrictive immigration policies (almost three-fourths of agricultural workers in America are immigrants, by the way), mean the farm can't harvest both crops.[4] Either the cherries or the peaches rot in the field.

That Bu knows any of this is incredible. Not a single buyer affiliated with concentrated animal feeding operations knows my father's or brother's name, for instance, let alone the ranch's history or current struggles. The relationship is one-way: the beef packers prefer certain breeds of cattle (which may or may not be suited to the rancher's environment), determine the weight the cattle should be at auction time, collude with other companies to fix prices, and pay zero premiums for regenerative practices. Whether my dad or brother stays in business makes no difference so long as cattle that meet industry standards keep showing up at the sale barn. That's not how VV does business. "Communication is key in a value chain where you really want your partners to survive and thrive," Bu says. "In a long supply chain, you don't have a story. You're not a person."

And as VV proves, people matter very much.

Days before I visit VV, the leaked Supreme Court draft decision about *Roe v. Wade* hits the news. I'm feeling low when I board the plane to San Francisco, though not as low as I will feel the next month when the ruling becomes official and the court asserts itself as far right. I'm a woman writer traveling alone, with limited grant funds I grappled to secure and no book advance, working on a project about female leadership in a male-dominated food system. Moving freely, controlling my body, building a career as a woman—all of it feels uncertain now, like the world is shifting into a version of *The Handmaid's Tale*. I wonder whether the book I am writing will be allowed a place in that world.

The ride share driver drops me off in front of the towering VV building, its green and yellow paint glowing in the morning sun as if lit from the inside. "Veritable Vegetable" and the company's circular logo are emblazoned, huge, on a side wall, with the motto "Delivering organic. Driving change" underneath. Here is something women created. Here is a physical manifestation of female ingenuity, leadership rooted in empathy, progress in the face of adversity. The building feels like a comforting monolith, a beacon of hope. The part of me feeling low needs to witness this kind of strength right now, I realize as I ring the door buzzer. I need to hear from women who thrived in harsher circumstances than my generation faces. Mary Jane, Karen, and Bu built something that likely seemed impossible to outsiders in VV's early days: an economically sustainable, women-led food distribution company that enacts environmental and social values instead of chasing profit and size, all in a capitalist economy and patriarchal society.

When the company restructured in 1988, Bu and Mary Jane became co-owners, with Karen joining the ownership team in 1995. Currently, Mary Jane is CEO, Bu is director of purchasing, and Karen is director of sales. The four of us gather in VV's conference room. Across the street, the company's state-of-the-art warehouse hums with activity I will observe on a tour later that afternoon. Trucks rumble past and we keep working on the cherries. I want to understand how VV has accomplished everything it has and how its model might be replicated elsewhere in the food system. "I know it sounds naive, but I think actually one of the fundamentals is setting aside making big profit. I really feel like we have worked steadily and consciously and intentionally over time to understand how we can have a good facility, good tools, pay a good wage, and enable our farmers to get a good return and our retailers to have a good price," Mary Jane says. "To me, if that's all in balance, everybody can have an appropriate part of the monies that are produced."

The women's outlook on profit reminds me of a passage by Robin Wall Kimmerer, scientist and member of the Citizen Potawatomi Nation. In *Braiding Sweetgrass: Indigenous Wisdom, Scientific Knowledge, and the Teachings of Plants*, Kimmerer reminds us that "for the greater part of human history, and in places in the world today, common resources were the rule. But some invented a different story, a social construct in which everything is a commodity to be bought and sold. The market economy story has spread like wildfire, with uneven results for human well-being and devastation for the natural world. But it is just a story we have told ourselves and we are free to tell another, to reclaim the old one."[5] Kimmerer's assertion that the market economy is a story of human existence that we can replace with an older, more equitable and ecological version is proven in VV. In their business model, food looks more like a common resource with benefits for all that Kimmerer describes.

At VV, the compensation ratio between highest-paid and lowest-paid employees is 4:1. Among the 350 largest publicly traded U.S. firms, the average CEO to typical worker compensation ratio is 351:1.[6] To put those numbers in context, the CEO-to-worker pay ratio was 21:1 in 1965 and 61:1 in 1989. Instead of fueling inflated salaries for leadership, profits at VV cycle into the business, toward philanthropy and green initiatives, and back to workers through benefits and to farmers through fair prices. For proof: the company is a certified B Corporation, certified California Green Business Innovator, San Francisco Legacy Business, USDA Good Agricultural Practices and Good Handling Practices certified, and has an organic handling certification through California Certified Organic Farmers.

Not that VV hasn't used profits to grow over its forty-nine-year history. Today the company has one-hundred-plus employees, a "green fleet" of thirty-five near-zero-emissions trucks, ownership of a 24,000-square-foot office and warehouse building, and a main 30,000-foot leased warehouse they remodeled in 2016. VV developed a robust composting program that together with other initiatives diverts 99 percent of company waste from landfills, and solar roof panels that offset 25 percent of its energy use. The company has earned numerous sustainability, transportation, and leadership awards over the years from entities like the Ecological Farming Association, the Organic Trade Association, the San Francisco Chamber of Commerce, Amy's Kitchen, *FleetOwner, Food Logistics, Heavy Duty Trucking*, and others. Mary Jane, Karen, and Bu managed all that growth carefully, resisting pressure to expand too quickly or scale up in ways that would hurt farmers and employees. "This is really trying to be intentional about what surpluses are created and how that gets put back in. Considering the entirety of everybody doing the work and the impact of the work that we're doing. A whole system approach," Mary Jane explains.

"It's not just growth for the sake of growth," Karen says. "You can reach a place where you can be a successful business without feeling like you have to continually grow, continually swallow up other companies or be swallowed up."

"Nature doesn't grow that way," Bu adds. "That's cancer, right, unlimited growth? That's the broader economic model we're in. It's not sustainable."

"Most successful communities in nature, they grow in complexity, but not in size, and that's what supports all the various components of those communities," Mary Jane agrees. "To the extent that we can compare ourselves to natural systems, many solutions to problems in the natural world are applicable to the problems that we see in the modern world."

Instead of expanding into faraway markets—growth for the sake of growth—VV invites like-minded people to learn from their model so those folks can improve an existing distribution business or start a new one. "We're big proponents of regional and community control and design, economic autonomy for communities," Bu explained earlier. "People have approached us over the years and asked, 'Why don't you have a VV in LA? Why don't you have a VV in North Carolina?' We've always said, 'Hey, you're welcome to come learn.' We've been mentors for many in what's considered the food hub community. We're the grandmothers of the food hub movement," she joked. Female mentorship and collaboration—those were rare experiences for Mary Jane, Karen, and Bu as they built the business because few women held visible leadership positions in the distribution world. They mention several influential female advisers and contemporaries, but the status quo of male control persisted. "As women, we had no institutional support," Bu explained. "We had few mentors, lack of access to capital, we had to figure things out for ourselves. Women as a potential economic force were not regarded. There were no Small Business Administration programs, for example, and no economic development programs that targeted women as a minority group or underfunded group."

Female representation in the food industry has improved over time but is far from equal. A recent study of food manufacturers, distributors, and operators found that women make up 49 percent

of entry-level employees but only 23 percent of C-suite executive positions (numbers that roughly mirror female representation
across all industries, not just food).[7] Of the three sectors, distributors have the lowest female representation, with 35 percent of
entry-level employees identifying as female and just 9 percent of
C-suite executives. Mary Jane, Karen, and Bu confirm these numbers anecdotally when I ask whether they have noticed any increase
in the presence of women leaders in their industry. "I see more
headlines in the produce community publications of women being
promoted into higher positions within produce industry companies. I see that for sure," Karen says. "But I don't necessarily feel
touched by it in any way. In our immediate surroundings and the
companies that we overlap with and are aware of, we haven't seen
any change in leadership with more women."

Mary Jane notes that often women *are* present within the food
industry, but they may go unseen. "As we observed coming up,
there was always, almost without exception, a woman's presence
on the farm," she remembers. "There were always women in any
company that was developing the organic trade, on the farm, in
retail or distribution; women were helping make things happen, but
largely behind the scenes." Bu describes any increase as "here and
there" and names a handful of food businesses in the region with
female leaders. "I do think that in the produce world I've seen an
effort in the past five, ten years to recognize that it was completely
dominated by older white men, and it needed to change," Bu says.

Not so at VV, where women hold all C-suite positions. "We
didn't intend to be women-led, but we were small, and when the
male founders left the company, we said, 'Hey, this is different, let's
keep it like this for now,' and that was in 1979," Bu said earlier.
Today a group of fifteen female department heads, directors, and
managers form the company's core leadership team, which meets

regularly about financial, operational, and other matters. About 45 percent of all staff are women, and the company intentionally develops internal talent for promotions, rather than hiring externally. Still, patriarchal attitudes crop up, such as when VV's warehouse operations manager, Jess Wilks, is still asked "Where is the manager?" by people entering the warehouse, or when director of finance Casey Castro felt pressured by male vendors to accept nonstandard payment terms. "It's still hard," Mary Jane acknowledges. "We do observe and experience that the male voice will tend to have authority at the table, across gender. People tend to listen to the male voice for whatever reason. Sometimes it's only because it's a male voice."

Female representation isn't just important for equality and making business more empathetic and people-centered. It's necessary for building a regenerative food system writ large. Women tend to think holistically about the relationships between health, nutritious food, vibrant communities, fair working conditions, and the environment—exactly the mindset necessary for creating regenerative foodways that prioritize all those concerns. I talked about this with Jessica Hulse Dillon, director of the Soil & Climate Alliance. Hulse Dillon's work involves facilitating large-scale shifts to regeneratively grown products within supply chains. That includes helping farmers transition to regenerative production and connecting them to regional buyers. Women are playing a crucial role in building out the missing middle of the supply chain to absorb regeneratively grown outputs, she told me. "A lot of the innovators we work with in companies, they happen to be women," Hulse Dillon says. "I'm not saying that some of them aren't men. I work with some great guys. But a lot of the real leading voices and the people that are really just pushing and pushing and pushing on this tend to be female."

Global warming makes that push for regenerative all the more urgent, since climate-related changes are already occurring within food and agriculture. Take what is happening along the 100th meridian, the approximate historical boundary between the wet, fertile East and the arid, less fertile West. The change from one environment to the other is gradual, not instant, but it is nonetheless apparent in the ecological differences on each side. Scientists analyzed decades of precipitation and evapotranspiration data and discovered that the arid-humid divide has shifted east to about the 98th meridian. Their models predict that line will continue crawling eastward as the planet warms because of climate change.[8] Farmers operating near that in-flux boundary tend to grow corn and soy, and yields are all but guaranteed to plummet in hotter, drier conditions given the region's weak soil. Even when scientists factor in technological advances, U.S. corn and soy yields are expected to drop in the face of higher temperatures and less precipitation.[9] Supply reductions are bound to occur as a result if farmers fail to adapt by changing their practices. They also risk going out of business, which could reduce food supplies and diminish rural communities even further.

Other climate zones are shifting around the world as well. The tropics are expanding, which means hotter, drier conditions farther northward and southward. The Sahara is 30 percent larger than it was in 1920. In parts of Canada, the permafrost line drifted eighty miles north in fifty years. At the same time, U.S. plant hardiness zones are moving northward at thirteen miles per decade. Perhaps most concerning from a food perspective is that the world's wheat belt is shifting poleward up to 160 miles per decade.[10] Farmers in parts of the U.S. wheat belt like Oklahoma and Kansas already feel this shift as crops fail due to heat stress. Some people believe the world can simply transfer wheat production north, but the reality, according to David Wallace-Wells in *The Uninhabitable Earth: Life After Warming*, is that "you can't easily move croplands north a few hundred miles . . . Yields in places like remote areas of Canada and Russia, even if they warmed by a few degrees, would be limited by the quality of the soil there. . . . The lands that are fertile are the ones we are already using, and the climate is changing much too fast to wait for the northern soil to catch up."[11]

At the same time, global and national food chains are inherently unstable in a warming world because of their concentration. In contrast, regional chains that source from a broad pool of farms and include a diverse network of distributors, processors, retailers, and so forth *are* stable; they are less concentrated and thus more flexible in times of change. As researchers have found, "Similar to the benefits of diversity in cropping systems for risk management, diversifying distribution networks has the potential to improve the stability of food availability when disruptions occur."[12] VV offers a glimpse of what a greener, more resilient, and more equitable regional food system might look like during times of crisis, climate or otherwise.

"For us, when the pandemic hit, it was like, 'Wow, we are really strong. We're very resilient,'" Mary Jane says. "We were able to pivot, we were able to help. The part that felt really good was that we were able to help get food to people who are food insecure and we could do it at cost or below cost. Because we're small enough, it makes us agile. We knew how to take care of our staff. We didn't have problems with losing staff. They felt safe, they continued coming to work. When things fell off precipitously for a little while, we were able to just keep pulling people and keep paying them a full wage. So that enabled us to be there to continue getting to the farms and getting the food to the stores. We were able to weather the big drop and the big spike and now the ongoing uncertainty."

Building more regional and local foodways that source regenerative products seems like a tall order given the current reality of our nationalized supply chains anchored in industrial agriculture. Yet VV, their organic farmer-partners, and their customers prove that regionality works. "The pandemic pulled back the curtain on supply chains," Bu says. "Regional is resilient and flexible. For us, it's value chains, not supply chains. A web of relationships, not a pipeline." Economic, social, and environmental priorities—nothing is sacrificed in this web of relationships VV has created that mirror the complexity and collaboration of the natural world. The company and others like it scattered across the country point the way forward, offering a template for regional distribution that can meet our food needs and withstand shocks. How the regenerative philosophy manifests will differ depending on the region and industry—what makes sense for High Plains beef distribution may not apply to Northeast vegetables—but that's a strength: to each region and industry its own appropriate methods for supporting regenerative producers, creating quality jobs, and supplying customers with responsible products. And we need a bold cohort of diverse people to lead the way in constructing these new systems.

Perhaps the question is not whether people *can* fix the food system, but whether enough people have the *courage* to. As Mary Jane, Karen, and Bu approach retirement and succession planning, I think they can safely say that yes, they found that courage. And they hope other people will, too. "It's not enough to criticize the paradigm," Bu says. "You have to be the thing that you want to replace a paradigm with."

5

Amplitude

Wen-Jay Ying, Local Roots NYC

Amplitude: the maximum displacement of the water's surface above or below the still line.

Wind collides with Brooklyn's skyscrapers and rushes downward into the streets, a phenomenon I later learn is called the downdraft effect. Absolutely bone-chilling is what I call it in the moment. It's January in New York City and I am walking. I woke hours too early for my eleven a.m. meeting and decided to kill time by hoofing it north from my Gowanus hotel toward the Brooklyn Queens Expressway, which will also reduce the Uber fare to my destination in Greenpoint. But my winter gear and conviction that no cold can be worse than South Dakota cold are not enough. I duck into a CVS and warm up while pretending to look at umbrellas (I have one in my backpack), then walk some more before popping into a bodega and wandering the aisles until the clerk eyes me and I buy a bag of Tropical Skittles and leave.

I walk to where Tillary Street joins the expressway, right outside the New York City Police Department's 84th Precinct. My eyes water from the wind. I think about the day ahead: I will meet Wen-Jay Ying, founder and CEO of Local Roots NYC, at the warehouse that stores the produce, meat, dairy, pantry staples, and many other goods that her business distributes through its Harvest Club. We'll talk, hopefully eat lunch—the walk has me hungry already—and end up in Carroll Gardens at Local Roots Market and Café, Wen-Jay's retail market/Chinese-inspired eatery, where I will meet some of her customers picking up their Harvest Club food shares. The day is all about food, regenerative agriculture, and connecting consumers and farmers—but waiting for my Uber in the nation's largest city, I feel as far away from a farm as one can be.

Thankfully, that feeling evaporates when I enter the warehouse. Local Roots' sourcing and warehouse manager leads me through a walk-in cooler, opening cases of bok choy, daikon radishes, acorn squash, Asian pears, and apples so I can snap pictures. Boxes filled with customers' orders (called shares) stand ready for delivery, the cardboard decorated with pink artwork of a turnip, an apple, a leek, and a steak along with the slogan "Kick back, relax, we bring the farm to you." I check out the specialty cheeses, smoked whitefish salad, and eggs. We tour a freezer stocked with various meats, tempeh, ice cream, and edamame dumplings. I peruse shelves of pantry items made with local ingredients and/or processed locally—apple cider vinegar from New York apples, soy sauce from New York soybeans, pasta with spent grains from Brooklyn Brewery, and salsa made with New Jersey tomatoes. One hundred percent of what Local Roots sells (except for wild-caught Alaskan salmon) comes from within a five-hour range of where I'm standing, and even as close as ten minutes away with urban farms.

The farm doesn't feel quite so far anymore.

Giving New Yorkers a connection to their food is a major reason Wen-Jay, thirty-eight, started Local Roots back in 2011, she explains when we sit in the upstairs office to talk. She is the epitome of a millennial business owner, a blend of casual and professional. Today she wears cream fleece sweatpants and a matching zip-up sweater over an olive tank top and an assortment of necklaces. She has straight bangs cut just above her eyebrows and long black hair that is loose except for a slender braid at the nape of her neck. Thin blue liner rims her eyes. Her way of speaking is calm, thoughtful, direct. To me she projects a distinctly Brooklyn vibe that is both trendy and down-to-earth, fun but also no-nonsense.

Before opening Local Roots, Wen-Jay read an article about food deserts in New York City. Without grocery stores close by, residents turned to bodegas and convenience stores that typically sell processed items and limited fresh food, all at high prices compared to grocers. These customers, who tended to be low-income, often developed health problems attributable to poor nutrition. With few resources available for health care, they returned to the same bodegas and convenience stores for medicine that masked problems rather than restoring their health. "I just felt like in this city that I love so much, it's this vicious cycle of food and poor health, and how a city of abundance can have so little in terms of easy access to good-quality food," Wen-Jay says. "That was my first real push towards this world of local agriculture."

Local agriculture is not exactly the world Wen-Jay imagined she would inhabit after graduating from Boston University with a psychology degree and moving to Brooklyn. After years of chasing academic goals, Wen-Jay craved variety and experimentation. She took a job at a clothing store, then at a luggage store. She played in five different bands and considered careers in affordable housing or social work. The idea of owning a business fluttered at the edges of her thinking. "Ever since I was a kid, I wanted to do something in my life that helped other people. I'm also kind of creative, so owning a business fills in a little bit of the creative bucket," she says. Wen-Jay is also a people person: "I loved the idea of being a store owner; it's like welcoming people into your home," she told me earlier.

These different impulses coalesced after a conversation with Wayne Coyne, lead singer of the band The Flaming Lips. Wen-Jay met Coyne backstage after a concert—she'd been selected to dance onstage in a Pink Power Ranger costume as part of his show, an opportunity she describes as "her dream"—and he asked about her life plans. Still trying on various careers and searching for altruistic outlets, she mentioned joining Hurricane Katrina reconstruction efforts. Coyne responded with life-changing advice, Wen-Jay tells me. "He said, 'That's really great, but I feel like so often we're always going to different places and following the next natural disaster. You should also think about what your direct community needs.' That really hit me hard," Wen-Jay says. Not long after, the article about New York City's food deserts landed in her inbox.

And so the journey began. Wen-Jay enrolled in the AmeriCorps VISTA program, which paired her with Just Food, a Brooklyn nonprofit that gives food-insecure New Yorkers access to afford-able, healthy food and offers community food training. She helped The Piggery, a family-owned butcher shop, facilitate their meat CSA. At the Union Square Greenmarket she sold bread from Hot Bread Kitchen, an organization providing bakery jobs for immi-grant women. She supported the city's growing CSA movement and also joined Red Jacket Orchards, where she launched their fruit-only CSA and tended their farmers' market stalls. Along the way she met farmers and food purveyors, learned about ecological sustainability and the importance of local eating, and translated her knowledge to customers. But this food journey halted abruptly when Red Jacket restructured its business and laid off employees, including Wen-Jay.

That was the first job Wen-Jay felt truly, completely absorbed in; she thought she'd found her calling, only to have the door slammed in her face. In a moment of clarity, the kind born out of crisis, she realized what she *really* wanted to do: create a CSA program that sourced from multiple farms for its weekly shares that was less work for the farmer and more convenient for the New Yorker. Her model would be a twist on the traditional CSA in which customers pay an individual farm or ranch in advance for a season's or part of a season's worth of product, generally picked up weekly or biweekly. To Wen-Jay, that structure didn't fit the busy lifestyles of most New Yorkers, who may bristle at long commitments with little week-to-week flexibility and are accustomed to variety that a single farm usually can't provide. A "one-stop shop" seemed obvious—but no jobs existed in that line of work, Wen-Jay discovered. Frustrated, she found herself in tears on Brooklyn's picturesque Court Street, seeking career advice from her mother over the phone.

"My mom said, 'Well, why don't you start your own business?' I was like, 'I am twenty-five years old. I have no business experience. No one is going to trust me.' It's hard to trust a concept. If you imagine you're a farmer, your livelihood depends on somebody buying your product at a very small profit margin. There's not a lot of room for error. And a twenty-five-year-old kid comes up to you, you have no idea who they are, and they're like, 'Trust me to buy out X amount of product in this new business model.' Multi-farm CSAs weren't really a business yet. People weren't familiar with subscriptions at that point either. There were farm boxes, but a CSA like mine wasn't a thing then." She also had no outside startup funds, no business degree, and no job to carry her financially until the business took off—if such a novel idea could flourish at all in New York, one of the most competitive and expensive cities in America.

Outsiders might have bet Wen-Jay's project would fail. But they would have been wrong, because this young entrepreneur had everything she needed to invent a unique, thriving food business.

Years' worth of experience and personal relationships in the local farming and food scene, a millennial's tech savviness, a storyteller's gift for marketing, an artist's ability to think creatively, a mind tuned to logistics and problem-solving, and an unshakable drive to build a better food system for farmers and consumers—these traits empowered Wen-Jay to launch Local Roots NYC just a few months after that conversation with her mother. And she had one more defining strength: her Chinese heritage and her life experience as a Chinese American woman.

Wen-Jay grew up in Garden City, Long Island, but agriculture and food are in her DNA. Her great-grandfather owned a grocery store in China. Her grandfather helped shape Taiwan's agricultural system, and then expanded Taiwan's agricultural exports to Germany. When her grandfather moved to America, he baked and sold red bean buns in Manhattan's Chinatown. Wen-Jay sees Local Roots as an extension of her grandfather's legacy, of aspirations he couldn't quite realize as an immigrant but that she can today. Wen-Jay's parents, both children of Chinese immigrants, also primed her to run a business, even if that was not their intention. They pushed her to work hard and be goal oriented, which contributed enormously to Wen-Jay's tenacity in operating Local Roots. "The only reason why I was able to survive as a business for so long, and without any kind of investment money, is working all the time," she says. "I feel like that work ethic I was raised with from immigrant parents, and especially as an Asian American, is the reason why I was able to start a business without much support outside."

Her parents also cultivated a love for Chinese culture, especially food, that she expresses by encouraging the farmers she sources from to raise Asian vegetables and by prioritizing Chinese dishes at her café. And the family's Buddhist beliefs coincide with Wen-

Jay's commitment to regenerative agriculture. In Buddhism, she explains, food heals and generates life, which regenerative practices also do through greater nutrient density, better soil health, and stronger agricultural communities. Balance is paramount in Buddhism, a concept Wen-Jay says is akin to the ecological balance sought through regenerative production. Given that Wen-Jay's brother is a Buddhist monk, like generations of family before him, this connection between her Chinese heritage and her work in the food system is powerful.

Wen-Jay's identity as a Chinese American woman means that she sees the world, and can therefore build a business, in a distinctive way. Her personality also influences her work: "My experience as an Asian American woman growing up in a predominantly white American town, and also as a Libra, it's nurtured a personality type that has led me to something like this," she says. "I really want to make sure people are happy or content. Not in a way that I'm a pushover, but I'm always making sure people feel comfortable or that I can think about what their needs are and I'm conscious of that. I think there's a sense of community that sometimes I wonder if I get from just being Chinese. It's not an American thing, community; America is a very individualistic country. But when I went to China my first time, it felt different because you are looking out for each other over there. I think that's kind of ingrained in me, taking care of each other."

Reciprocity, community, and care. Wen-Jay's words remind me of Kelsey and how her Lakota identity informs her regenerative practices and business decisions at DX Beef. They also echo Carrie and Erin and how their Black heritage and love of community influence their work at Footprints in the Garden. Wen-Jay's assessment of the way her background shapes Local Roots is a reminder of why diversity is essential for the regenerative food movement.

We need the widest possible variety of perspectives for remaking a system dominated by *one* type of thinking: that cheaper, bigger, and homogenized is better when it comes to what we eat and how we grow, process, and distribute food, and that profit is the best indicator of success.

Not that Local Roots isn't profitable. Like any business, it has to be. "To me, the best way to prove that this can work and make a difference in the world is to prove that it works as a business and it doesn't need investment money or doesn't need nonprofit funding," Wen-Jay says. "You can't keep supporting the farms you want to support if you're not able to pay yourself or your staff or keep your business going or keep growing. The biggest impact you can have is actually to be a profitable business and help everyone around you." What that looks like is steady, manageable growth paired with values-based decisions. Since inception the business grew to include the Brooklyn storefront/café as well as two kiosks in Manhattan. The Local Roots staff increased over the years from two to twenty people, most of them women. "A lot of women want to work in this space, so it's easier to hire them," Wen-Jay says. "It's also a little conscious. Of course I would like to give more women opportunities, especially women of color. I'm obviously not going to choose one person or the other based on their gender or their ethnicity. But I think we attract more people that are diverse just because of who I am." That's another reason the food system needs women and women of color in particular: their presence emboldens others to participate.

So how does the Local Roots CSA model work exactly? I asked Wen-Jay. Harvest Club customers sign up online for a weekly subscription based on their taste, budget, and household size, she explains. There's the Basics Bundle, for example, which includes a dozen eggs, a quarter pound of salad mix, and the choice of a half pound of meat or mushrooms. Then there's the Essentials Bundle, which comes with three types of vegetables, half a dozen eggs, two pounds or pints of fruit, and two packages of meat. Larger subscriptions are available as well. Customers could also select subscriptions just for milk, meat, cheese, or bread, or combine multiple subscriptions. They can add dozens of items à la carte week by week, like herbs, pasta, sustainably caught seafood, meat substitutes, crackers, salsa, jam, spices, sauces, kombucha, even candy. "Our CSA model is unique in that we really try to provide as much variety as possible to the consumer, so that they don't get tired of the CSA," Wen-Jay says. "[Having many vendors] is important in terms of variety, for preference for the consumer, but also for nutrition, to diversify the nutrition we're able to provide to people."

Wen-Jay sources from more than fifteen farms and twenty food makers. She chooses whom to work with based on her belief that food production should improve soil health and carbon sequestration, use water responsibly, prioritize animal welfare, give farmers a decent living, and contribute positively to local economies—all of which Wen-Jay verifies through farm visits and conversations with producers. Because she sources from so many places, her business is categorized as a type of multi-farm CSA with add-on options. In her CSA design, a nonfarm entity (Local Roots) aggregates, packs, distributes, and markets the shares; in other multi-farm CSAs, farms may band together via a cooperative or other arrangement and handle all things share related themselves.

Researchers have named seven additional types of CSAs ranging from the traditional (members pay one farm at the beginning of the year and enjoy shares of whatever is harvested) to market style (like a seasonal box, but products are displayed at a distribution site and customers pack their share as if they are shopping).[1] Shares from a multi-farm CSA boast variety and flexibility if something goes wrong. If one farmer's cauliflower crop fails, for instance, then another farmer likely has some. Rather than asking farmers to commit to producing set amounts each week, Wen-Jay promises to buy whatever they have, which she says many farmers prefer. "A lot of farms, after I would talk to them more about this concept, it's a lot of pressure for them to grow certain quantities for that many varieties every single week for someone," she explains. She and her staff also plan with farmers to grow specialty and Asian produce to increase selection and better match the tastes of diverse customers. Wen-Jay says farmers are more willing to experiment with unfamiliar crops because they have a guaranteed customer: her. Farmers drop off their product at the warehouse and the Local Roots staff packs it to order or distributes it to Local Roots Market and Café. Customers pick up their orders at locations around the city or opt for delivery. They pay weekly and can skip, cancel, or modify orders as they please.

This kind of choice and flexibility is incredibly important for a CSA to survive. Daniel Prial, agriculture specialist with the National Center for Appropriate Technology, writes that while national CSA membership is increasing, customer retention rates are only about 45 percent. Lack of choice is the main reason customers don't stick with a CSA. Prial argues that farms need to diversify, get creative, and offer customers more product options and control over their shares if they want their CSA to thrive, especially as national companies like Blue Apron or FreshDirect take market share away from CSAs.[2] CSAs also may flounder if farmers lack the bandwidth and financial resources to run them well. Many producers are excellent at growing, but they may not be as comfortable or skilled with operating a website, understanding purchasing software, or marketing to the public. Small businesses like Local Roots can provide that expertise and time investment instead. It's like a partnership. "Another reason why I started Local Roots is that I want to take those operational logistics away from farmers so they can spend more time farming," Wen-Jay says. "I know how much work it takes to do education, to do logistics, do all those deliveries and all that kind of stuff. This is my full-time job, plus other people's full-time jobs at Local Roots."

In fact, it's more than full-time for Wen-Jay, who tends to work ten-plus-hour days, like many small business owners. She admits to being "not very good at schedules" but tries to split her time between the Harvest Club, the café, and marketing or other public-facing work. "I would love to find a way where it doesn't take up my entire life," she says. The hours she puts in now, though, pale in comparison to the long days in the most difficult period of her professional life.

It was late December 2019. Wen-Jay was in Shanghai with her parents and brother, all Texas residents. They were visiting Wen-Jay's other brother, who resides in Shanghai with his family, which had just welcomed a son. Wen-Jay, her parents, and her siblings convene about every five years, so the trip was a long-awaited, joyous occasion—until news reports began circulating about a strange and deadly illness. The family planned to follow their stay in China with time in Japan, a detour Wen-Jay typically schedules when she is in Asia. Japan seemed safer from COVID-19, as the illness had been named, so the family continued with their post-China plans. In hindsight, Wen-Jay marvels that everyone arrived home healthy.

The frightening news reports repeated themselves on U.S. televisions not long after Wen-Jay's return. She understood the situation better than most Americans, certainly better than I did watching those same news stories in North Carolina. When lockdown began in early March, Wen-Jay's staff implored her to shutter

Local Roots, if only for a few weeks. But she sensed the irrevocable harm that could follow if she did. "One, few small businesses can take a break and survive, especially with groceries, but I think with any business, consistency is really important to customers. Customer loyalty is also really hard to get, so once your customers go somewhere else, they are going to stay there," she explains. "So I knew [shutting down] wasn't possible, and I also had a bad feeling because I had seen how long the virus was actually out in the world, longer than most Americans knew. I knew there was no way this was just going to be a week because I saw this on the news two months ago."

Shutting down could diminish Local Roots, stop the flow of income to employees and farmer-partners, and leave customers stranded as grocery store shelves emptied. Staying open could mean illness or even death. Wen-Jay had to decide carefully but quickly. The smartest way to help people around her and protect Local Roots, she concluded, was to stay operational. "Even if I didn't know what the turnout would be as a business, at least I could feel like I was helpful to the people in my community during this time," she says. Almost overnight, she pivoted Local Roots from mostly pickup to mostly delivery, which involved buying a new software program for order management, hiring staff, and finding vehicles. "We might have had fifteen households, tops, in our delivery pool, maybe twenty-five, before the pandemic. Within two days we had over five hundred," Wen-Jay says. "It's good that I like fixing problems and figuring out logistics, because I was like, 'Okay, we have to hire ten times more employees.' Between the warehouse and getting drivers, we eventually found twenty-ish people within two days." As other businesses laid people off, Wen-Jay was hiring.

Wen-Jay also expanded her farmer network to meet the surge in demand. Remember, it was early springtime; the abundance of summer harvest was months away and over-winter storage crop inventory was low. She bought whatever farmers and food purveyors had in stock, even the unfamiliar. "I do remember asking our farmers, 'Is there anything that is growing on your farm that you don't think can sell but is actually edible, like sweet potato leaves, things that maybe go to waste in a farm, but we can introduce to people?' I think we did a really good job with the supply chain. And I think our farmers did a great job with that, too, compared to those larger industrial farms that weren't able to pivot as quickly," she says.

Where nationalized food chains fell apart, shorter chains experienced stress but stayed intact. Consumers turned to local and regional outlets, as shown in an analysis conducted by a team of experts assembled by the USDA Agricultural Marketing Service. The team surveyed five thousand residents in non-metro, medium-metro, and large-metro communities about their food expenditures in eight distinct categories, like supermarkets, farmers' markets, food boxes, and independent local restaurants. They asked about spending in three time periods: September 2019, April 2020, and September 2020. The survey found that "following the initial shock of COVID-19, residents in many different communities increased expenditures at market channels where they can find local or regional foods. In particular, farmers markets and food boxes have become more prominent" across the board. In September 2020, food box spending was 30 percent higher in small communities, 17 percent higher in medium communities, and 14 percent higher in large communities compared to September 2019.[3] Another Agricultural Marketing Service study found that food box spending continued to climb from October 2020 to October 2021 in medium-metro areas, as did farmers' market spending in large-metro and non-metro communities.[4]

These outcomes suggest that people look to local and regional food sources in times of crisis—that they *need* those sources. People likely do the same in the aftermath of climate-driven emergencies that interrupt national and global food chains. That is why we need to strengthen and expand local and regional food chains founded on flexible, diverse farms and food makers. Such foodways are critical for ensuring stable food supplies during the ever-increasing natural disasters we are experiencing with climate change.

Wen-Jay faced multiple challenges by staying open, from keeping workers safe from an ever-changing virus to complying with evolving public safety regulations. "There was never stability," Wen-Jay recalls. "Every day felt like it was a whole new pandemic to figure out for months on end every single day." Thankfully, the challenges were a bit easier to handle because of Local Roots' small size and centralized leadership. Wen-Jay staggered employee schedules without sacrificing productivity and instructed anyone whose job allowed remote work to stay home. As a result, COVID outbreaks did not slow or halt orders at her warehouse like they did in industrial meat-processing plants and other workplaces where employees toiled in proximity. She was able to donate hundreds of pounds of food to essential workers during the pandemic, too.

Like the independent farmers she sources from, who scaled up and responded to changing demand more seamlessly than large-scale conventional farmers did, Wen-Jay implemented changes rapidly when needed without navigating layers of corporate red tape. For example, she recalls one pandemic-era delivery when a brownstone collapsed and blocked a street, and a driver could not reach a customer's home. "I live in the area, so I got the delivery box and texted the customer and walked it over to them," Wen-Jay recalls. "The customer was like, 'This is my first time leaving my house in months.' Because she has a kid who has a weakened immune system. It's interesting when you see those little inside moments of people's lives during the pandemic and how integral we were in people's lives and how this got them through everything. That was a point of mine to make sure that anyone who wanted food from us would always get it."

Wen-Jay's delivery of that food box, and the way she, her staff, and her farmer-partners adjusted to a changing economy and working conditions during the pandemic, illustrates the importance of local control within the food system. Local control means faster adaptive responses when the unexpected happens. Local control also leads to more variety in those adaptations. How Local Roots transformed looks different from how other food businesses changed during the pandemic, not only in Brooklyn, but also across the United States—and that is a good thing, because if some businesses fail in their response, then others fill the gap to help prevent all-out failure, a concept called redundancy or overlapping systems. Redundancy fosters the resilience we need in times of both crisis and stability.

But over the last hundred years, our food system abandoned overlapping systems and diversity in favor of centralization, standardization, efficiency, and cheapness, explains Mary K. Hendrickson, associate professor of rural sociology at the University of Missouri. "Decision-making has to be decentralized in order for us to really employ all the adaptive mechanisms that we can," Hendrickson tells me over Zoom. "That is not going to happen if you have big farms, big consolidated food chains that are only focused on efficiency. It hasn't allowed for redundancy, and redundancy is going to be really important as we cope with climate change. We want a lot of backup systems, and right now when we've consolidated, we have only thought about efficiency and we haven't thought about redundancy and all the different kinds of backup systems we might want to have. And that's important for food safety. That's important for climate change, it's important for most of the things that we're dealing with. I think really we need to think about control, as much control being exercised as possible at the community level. It's both a bottom-up and top-down kind of approach that we need."

Three Princeton University researchers considered the milk industry as an example of the need for decentralization. During the pandemic, massive dairy-processing companies struggled to redirect milk intended for places like restaurants and cafeterias to retail outlets and consumers who needed it. And the huge dairies supplying that milk failed to pivot to other buyers, mainly because the industry is so centralized and those dairies are too big to move milk into local channels instead. Milk is a time-sensitive product; cows can't stop producing it at a moment's notice so their owners can develop new supply chains. So dairies dumped their milk and eventually began donating thousands of gallons to food banks, all while shoppers encountered shortages in stores. "Milk waste and donations are signs that supply chains lack resilience—the ability to bounce back from stresses, like a rubber band that returns to its normal shape after being stretched," the Princeton team writes. They go on to assert, "The fact that the U.S. has too much milk for some places and too little for others highlights weaknesses of conventional food supply chains amid shocks like COVID-19."[5]

The Princeton researchers argue for "food systems that are flexible and diverse," and they point to New Jersey's dairy farmers and their pandemic experience as a case study. Most New Jersey dairies are small and independent. Instead of selling milk to corporations or big cooperatives, they convert it into value-added products—cheese, yogurt, bottled milk—that they sell on the farm or through local retailers and restaurants. Others may sell raw milk for surrounding businesses to process. These dairy producers partner up via small cooperatives and work with other types of farmers and small businesses. For example, some dairies also operate slaughterhouses that process animals from neighboring ranches. Some own cafés or markets that aggregate milk, produce, and other goods sourced nearby. This system in which food flows directly or almost directly from farm to consumer is efficient and adaptable, even during a global pandemic.

Farm cooperatives like the ones in New Jersey are not a new invention, by the way. Co-ops exist across agricultural sectors, from grain and livestock to produce and sugar. Some buy and market their members' products; others provide cost-shared inputs and financial resources like loans and insurance. Dairy co-ops are the most common and largest type of co-op in America—so large that many act more like corporations than farmer cooperatives.[6] Researchers with the Open Markets Institute write that "milk producers historically organized into co-ops to counterbalance concentrated power among buyers, but today, the largest co-ops, such as Dairy Farmers of America and Land O'Lakes, prey off of small-scale producers they're supposed to protect and strike arrangements with massive milk processors, like Dean Foods."[7]

The grassroots New Jersey co-ops analyzed by the Princeton researchers, however, set their own prices, share resources, and generate profit for all. This agricultural economy works for everyone involved and is adaptable thanks to local power and partnerships. "New Jersey's local farms are able to bounce back from disturbances like a pandemic because they add a collaborative, 'horizontal' element to vertically structured farms," the Princeton team writes. "As networks of farmers and consumers grow, they become more connected and are able to flexibly pivot and adapt to meet demand, thus creating increasingly resilient regional mosaics of farms and customers."[8] Such decentralized systems could help protect our food supply and agricultural communities from climate change–driven disturbances. These systems will not stop a disaster, of course, but they can remain stronger and recover more quickly than their centralized, large-scale counterparts.

To me, the idea of local control also feels empowering for farmers and ranchers. I imagine the dairy farmers dumping their milk in 2020 felt helpless, choice-less, the opposite of in control and adaptable. After selling milk into a concentrated market for decades, how could they possibly reorient their entire operation and make connections with local and regional consumers and businesses? Decades of economic and social pressure to "get big or get out" boxed many conventional farmers and ranchers into expansive, highly specialized business models, which makes change difficult. I empathize with producers—I come from a conventional agribusiness family—and I know most have noble intentions. But after seeing the food and agriculture failures during the pandemic, I am scared for these producers' future if they do not embrace regenerative practices that will help prepare them for the disruptions of a changing climate. The pandemic was a warning, not just about the food system but also about fragility within our health care, education, manufacturing, and other systems. The question is whether society will learn and adapt, or ignore and risk failure.

Wen-Jay and I eat lunch at a nearby restaurant—she orders buttermilk pancakes with blueberry compote—and then we ride the subway to Carroll Gardens. A short walk brings us to Local Roots Market and Café on Court Street. The day hasn't warmed much and the wind is still brisk, facts that Wen-Jay, who hates the cold, remarks on with annoyance. The café is delightfully warm inside, though. Immediately Wen-Jay shifts into hospitality mode, jumping in to answer customer questions about bubble tea. She inputs their order and prepares the teas, and invites them to the Lunar New Year party at the café this weekend. Meanwhile, I browse the market shelves and coolers stocked with items I recognize from the warehouse: bagels, oat milk, meat, butter, beer, kombucha, spices, dry pasta and grains in self-serve bins, all manner of jarred goods, and house-made food like overnight oats and tea eggs. Loose vegetables and fruit for purchase line the café's front counter. A small side room is devoted to organic CBD, sustainable beauty products, and Local Roots merchandise. I scan the menu and wish I had room for a bao bun or scallion pancake. "These dishes are some of our founder's childhood favorites," the menu reads. "90% of our menu ingredients are sourced from small, local farms."

A few minutes later, an employee assembles the CSA produce in stainless steel tubs and decorative market-style piles, creating an attractive pickup station. This week it's acorn squash, red Russian kale, rainbow carrots, bok choy, black radishes, Asian pears, and cameo apples. A sign denotes what farm each product came from. Harvest Club subscribers appear with reusable bags when the pickup window opens. The employee distributes what's included in each person's share, answering questions and inviting people to handpick items that don't require weighing, like a bunch of kale or an acorn squash. Many customers have elected to add salad mix, meat, mushrooms, and eggs. One woman describes how the previous week's mushroom variety soaked up more sauce than other mushrooms she's tried—feedback Wen-Jay can keep in mind and relay to farmers. The customers talk with one another in line. The scene strikes me as a moment of community and connection in a big, busy city. And on the opposite end of these CSA pickups, local farmers, ranchers, and food makers are earning a decent living, reducing carbon emissions, delivering products with higher nutrient density than most conventional equivalents, and nurturing ecosystems below- and aboveground.

Whether consumers make the nutrition–environment–local economy connection isn't always clear, though. People are more educated about these topics now than in generations past, but Wen-Jay still feels like expanding her customer base is an uphill battle, in part because consumers lack knowledge about the food system and why local eating matters. I get it; I've spent years studying the food system and still feel underinformed. And Big Food isn't exactly forthcoming to consumers about farming and processing, or the environmental and social consequences of industrial eating. Wen-Jay sees it as her mission—her responsibility even—to convey information through social media, speaking engagements, face-to-face conversations, and other creative means. "Just because people are not buying local or buying organic doesn't mean they don't care. It might be that they just don't know," Wen-Jay says. Plus, psychological research confirms that humans find even small changes difficult, she points out. That includes buying and eating habits. "It's going to take a reminder every single day," she says. "You just keep feeding them a little bit of information constantly, until one day they might buy a carton of eggs from a local farm that's pasture-raised and that's what they stick to."

Wen-Jay believes high-level actors are necessary for a system-wide revolution. "Change has to happen on an institutional level. I love the home cook, I love the community of customers we have, and I know they're making a change because, you know, they might buy our things and they might convince their office to also," she says. "That has a ripple effect, but I do think large-scale purchasing is really where people like myself and other regenerative agriculture activists need to be focusing on. I'm still going to keep working towards my mission, but I also am not naive to how hard that actually is."

Turns out, a diverse collective of women are working on this issue, too.

6

Height

Tina Owens, Nutrient Density Alliance

*Height: the elevation of a wave from the trough (bottom)
to the crest (top).*

The Anaheim Convention Center in California boasts more than 1 million square feet of exhibit space. I'm not sure exactly how much of that is dedicated to food, beverage, and ingredient vendors at the 2023 Natural Products Expo West, but it is more square feet than I can walk in the time gaps between presentations and meetings during my three days there. I have never attended Expo West or its sister event, Expo East, before, or any such industry gathering besides the annual Association of Writers & Writing Programs conference. When I register for Expo West, I know turnout will be substantial. I see big-time names from big-time companies, organizations, and government entities on the speaker lineup. Even one of my writer idols, Bill McKibben, is slated to speak live-remote during the special Climate Day. I anticipate nauseating crowds and a good dose of brand propaganda. But ironically when I arrive at this mega-expo dedicated to food, what surprises me is the sheer quantity and diversity of fresh and packaged food products, supplements, beverages, ingredients, and more in the expansive exhibit halls.

Confession: I get a little carried away with the samples. I walk out of those halls every day (sometimes twice a day) swearing I won't say "yes, thank you" to quite so many offers to taste a bite of grilled steak, or a veggie gyoza, or a square of gluten-free pizza, or a tiny cup of organic coconut rice. I fill three tote bags with shelf-stable goodies. Peanut butter, fruit packets, candy, granola, sauces, cookies—you name it and I probably brought it home. Startup brands, medium-sized brands available in places like Whole Foods and Trader Joe's, even large companies like Annie's and Organic Valley jostle for the crowd's attention, inviting them to try the latest and greatest new product. I eat some incredible food. But honestly, I see mostly junk food, in other words, ultra-processed food (UPF), at this gathering of brands supposedly cranking out natural, healthful products.

UPFs contain added ingredients you might expect—sugar, salt, fat—as well as industrial, lab-created additives like chemical sweeteners, colors, and flavors, stabilizers, modified starches, fillers, and so on. Whatever you cannot easily recognize on an ingredients label is probably an ultra-processed additive. UPFs make up 57 percent of the American diet.[1] That is bad news for our health, as numerous studies show that UPF consumption is associated with weight gain, cardiovascular disease, coronary heart disease, cerebrovascular disease, high blood pressure, high cholesterol, some cancers, and higher mortality risk.[2] The UPF-focused Western diet also kills microbiota in the gut, known as our microbiome. These microbes are imperative for the immune system to function normally; without them, people can develop autoimmune diseases and many other ailments.[3] As infectious diseases doctor Chris van Tulleken argues broadly in *Ultra-Processed People: The Science Behind Food That Isn't Food*, UPFs are not actual food. "Addictive edible substances" is his way of describing the industrial creations people eat almost every day.

Walking the exhibit halls, I spot cereal bars with sweeping wellness claims, long lists of lab-derived ingredients, and eyebrow-raising calorie counts. Teas and juices with more "natural" sweeteners than a person should consume in a week. Plant-based "meat" chock-full of soy, salt, and mystery additives I need Google to help me understand (sorry, but Beyond Meat, Impossible, and the like are UPFs). "Organic junk foods are still junk foods," as renowned nutritionist and public health advocate Marion Nestle puts it.[4] An organic corn-based, cheese-flavored chip is preferable for avoiding chemical residues, GMOs, and questionable food fillers, but that chip is still a UPF that offers essentially nothing nutrition-wise.

One food expert calls UPFs "pre-digested" because the manipulation—refining, pounding, heating, melting, shaping, extruding, and additive packing—breaks down the food's internal matrix. That degraded food matrix results in fewer bioavailable nutrients. Processed food becomes so easy to digest that we eat more than necessary without feeling full.[5] And because the National Organic Program rules allow industrial practices like monoculture cropping and animal confinement, many Big Food organic products still cause environmental harm, albeit less overall than their conventional counterparts. As food journalist Mark Bittman writes, "If there was any promise of integrity in the mass-produced organic sector, it was destroyed. Most organic food is now part of the larger system, one that still fails to ask the key question: What do we grow food for?"[6]

As with organic, Big Food is flocking to the concept of regenerative agriculture. PepsiCo promised to implement regenerative agriculture on 7 million acres by 2030.[7] Walmart plans to ensure

North America as senior director, agricultural funding and communication. There she kicked off an on-farm return on investment tool intended to help farmers understand the financial benefits of regenerative, soil-centric practices. Tina delivered congressional testimony to the Select Committee on the Climate Crisis in Washington, DC, and began building one of the first-ever stacked financing models for regenerative transition that combined grants, impact investing, and philanthropy. In other words, she amassed government and other awards, investor funds, and philanthropic gifts and matched those dollars with farmers transitioning to regenerative agriculture across multiple regions. She continued that work as Danone's senior director, food and agriculture impact, and also concentrated on reducing scope 3 emissions within the company's dairy portfolio.

Today Tina works for the nonprofit Green America as senior fellow of regenerative agriculture at the Soil & Climate Alliance, co-leading the Nutrient Density Alliance. The Nutrient Density Alliance is a group of brands, nonprofits, researchers, and other entities acting collectively to advance regenerative agriculture by educating the public and fellow food system players about the relationship between regenerative practices, nutrient density, and consumer buy-in. "You see how far you can get within those systems and it's only ever for the good of one company," Tina explains. "I'm not interested in limiting myself to doing good only for the benefit of one company anymore. It's got to actually be for the full system." Since 2018, Tina has also run a small regenerative farm with her husband in Michigan. They raise heritage pigs on silvopasture, sheep in rotational grazing, free-range turkeys, geese, and more, all on farmland once used for extractive conventional agriculture.

I meet Tina in person for the first time at Expo West, where she is slated to participate in the expo's first-ever panel discussion on nutrient density and why brands ought to tell customers about the higher measurable nutrition in consumer packaged goods made with regeneratively grown ingredients. Representatives from Applegate, Patagonia Provisions, the Non-GMO Project, and SPINS (a food data company with a natural and organic emphasis) will contribute as well. The next day Tina will deliver a smaller off-site talk with similar themes. We convene at a nondescript table outside the Hilton Anaheim Food Court. Other meet and greets are happening around us, per the usual for a trade conference. Tina reminds me of the actress Hannah Waddingham, known for playing Rebecca Welton in the comedy series *Ted Lasso*. She is tall and blue-eyed with a warm smile and strong build. She looks as suited to speaking in a corporate boardroom as she is to wrangling livestock on the Michigan farm where she grew up.

Tina believes nutrient density is the next frontier for the regenerative movement. For a long time, she says, the conversation centered around soil health, climate readiness, and better financial outcomes for farmers—all important, but not necessarily to consumers who are being asked to pay more for regenerative products. Consumers need education about the relationship between nutrient density and regenerative agriculture in the form of peer-reviewed research (which already exists but could be more accessible) and in messaging on packaging, brand communication platforms, and the like. Armed with that information, Tina believes, they are more likely to choose regeneratively grown food.

"We are all chronically nutrient-deficient. Regenerative agriculture is able to bring that back in, in meaningful ways that are measurable. Consumers know none of this, and yet 66 percent of them are purchasing for health when they're in the grocery store and brands are not having this conversation with them," Tina told me in an earlier discussion over Zoom. "If consumers understand that it matters for their health, their fertility, their longevity, their kids' ADHD, their kids' emotional state and management, because we are all walking around nutrient-deficient, then I'm not being asked to buy something to save the farmer and the soil and the climate and the planet. I'm being asked to buy something to save myself. I am so much more interested as a consumer in saving myself," Tina reiterates today.

The human body requires macronutrients and micronutrients to function, but studies confirm that many Americans have nutrient inadequacies. For example, 45 percent of Americans lack enough vitamin A, 46 percent lack vitamin C, 95 percent lack vitamin D, 84 percent lack vitamin E, and 15 percent lack zinc.[10] The high-calorie, low-nutrition diet prevalent in this country can lead to "hidden hunger," in which the body suffers malnutrition and obesity simultaneously. A handful of conditions associated with nutrient inadequacies include cancer, osteoporosis, cardiovascular diseases, impaired immunity, general fatigue, and cognitive deficits.[11] In the United States, "poor nutrition causes more than half a million deaths each year, with disproportionate impact on Indigenous people, people of color, and low-income people in both urban and rural communities. Three-quarters of American adults are now either overweight or obese, and nearly half suffer from diabetes or pre-diabetes."[12]

Social inequities, poverty, food deserts, predatory marketing, and junk food are each to blame for inadequate nutrition, but so is the fact that the whole foods grown in this country are far less nutrient-dense than they used to be. According to retired University of Texas at Austin chemist and nutrition researcher Donald R. Davis, "Studies of historical nutrient content data for fruits and vegetables spanning 50 to 70 years show apparent median declines of 5% to 40% or more in minerals, vitamins, and protein in groups of foods, especially in vegetables."[13] The grains, seeds, produce, and other plant-based food crops we eat, including the processed foods built from them that dominate our diet, supply fewer nutrients per bite. As climate journalist David Wallace-Wells puts it, "Everything is becoming more like junk food."[14]

The cause of nutrient decline in food is threefold. First, there's the "dilution effect," or the inverse relationship between yield and nutrient density. A given area of soil only has so many nutrients available, so when yields rise, those nutrients are diluted among more crop output. Second, industrial farming practices like synthetic fertilizer applications and monoculture cropping that increase yields do so at the expense of soil health, which leaves fewer nutrients for plants to take up.[15] "Ultimately, the nutrition of the plant is from the soil, or at least from the root system," says Dr. Jill Clapperton, principal scientist and CEO of Rhizoterra Inc., and the founder of the online Global Food & Farm Community. "If we're going to get to the foundation of food, we need to get to the soil and we need to start understanding the nutrition of the soil. We know, just like ourselves and animals and everything else on this planet, that if nutrition is mediated through a biological system, it gets taken up better and more efficiently. It's the same with plants."

Third, because of human activity the planet's air contains more carbon dioxide than ever before, and as CO_2 levels have risen, plant mineral content has fallen. Plants need carbon dioxide to grow, but as biologist Elena Suglia writes in *Scientific American*, "Extra carbon dioxide acts like empty calories or 'junk food' for the plants, which gorge themselves on it to grow bigger and faster, consequently getting larger but less nutrient-packed."[16] That CO_2 junk food also supercharges weed growth.[17] CO_2's effect on nutrition will escalate if we fail to rein in greenhouse gas emissions. When scientists exposed crops like wheat, corn, soybeans, and field peas to carbon dioxide levels expected by 2050, they found significant declines in protein, zinc, and iron.[18]

More carbon dioxide also leads to sizable, ongoing quality declines in pasture forage. Researchers analyzed semiarid mixed-grass prairie over seven years and found that elevated CO_2 levels can increase the amount of forage but reduce its digestibility, which lowers livestock weight gain and milk production. More CO_2 also decreases plant nitrogen content, which results in less digestible protein (a source of amino acids) for livestock.[19] Given these associations, declines in forage quality could lead to less nutritious animal-based products. Studies are underway to quantify this loss, and there is already evidence that reduced grass nutrition due to elevated CO_2 levels causes die-offs in herbivore insects like grasshoppers.[20] A larger cause of suppressed nutrient density in meat, milk, eggs, and other animal-based products is the nutritionally depleted industrial grain rations livestock consume in concentrated animal feeding operations. Compared to meat and milk from feedlot-fed animals, meat and milk from pasture-fed animals contain more vitamins, trace minerals, conjugated linoleic acids, omega-3 fatty acids, and phytochemicals once presumed to exist only in plants.[21]

Some people argue that not eating animal products would solve at least some of our health and nutrition issues. Even after accounting for lifestyle factors like smoking, drinking, and obesity, researchers found that people who reported eating red and processed meat three or more times per week had higher risks of heart disease, pneumonia, diverticular disease, colon polyps, and diabetes. Poultry eaters increased their risks of gastroesophageal reflux disease, gastritis and duodenitis, diverticular disease, gallbladder disease, and diabetes.[22] There's abundant evidence that cardiovascular disease, type 2 diabetes, metabolic syndrome, obesity, and cancer risks decrease for people eating a Mediterranean diet, which features low meat and processed-food intake and high vegetable, fruit, olive oil, grain, legume, fish, and whole food intake.[23]

People should certainly limit animal products, mainly red and processed meat, for health reasons. But equally important is ensuring that the animal products we *do* eat are nutrient-dense because they come from the land, not from a feedlot or a lab. In reasonable proportions, unprocessed and sustainably raised animal products are part of a healthy diet, as Mediterranean diet advocates have shown. Chef Sean Sherman points out in *The Sioux Chef's Indigenous Kitchen* that wild game—as opposed to domestic pork and beef—is part of traditional, nutritious Indigenous diets that are hyperlocal, ultra-seasonal, incredibly diverse, and mostly plant-based. Other peoples, like the Kyrgyz in northern Afghanistan, the Evenk and Yakut in Siberia, and the Inuit in the Artic, lived healthfully on meat-based diets that reflected the limits and gifts of their environment. Only when processed Western food crept into those societies did illnesses like diabetes, heart disease, and hypertension appear.[24]

Eliminating processed food is probably the best move people can make for their health. It is probable that conventionally grown and/or processed food lacks biocompounds that we know support human health but do not currently measure in food, and also ones we do not know about or fully understand yet. Tina points out that society's concept of nutrition—that is, the components listed on a typical side panel like salt, sugar, calories, protein, vitamins, and fat—is grossly inadequate. Research confirms the presence of more than 26,000 distinct biochemicals in food, but the USDA and other nutritional databases track just 150, or about 1 percent, of those biochemicals. Scientists have called this food knowledge gap "nutritional dark matter" and are using machine learning to better understand the full biochemical spectrum of food and the health implications.[25] Scientists are also harnessing the technology that mapped the human genome and microbiome to uncover the molecular makeup of food and how that relates to health. "One of the first things I do with people when I want to help them understand how nutrition is an outcome of regenerative agriculture, I actually start with that 1 percent thing because you have to break people's religious level of faith—which I myself had at one point—that the side panel is nutrition," Tina says.

Educating consumers about nutrient density is critical for stimulating the market demand necessary for Big Food to stick to and expand regenerative agriculture commitments—promises initially made in response to climate change risks. Companies see the writing on the wall for conventional production on a hotter planet: soil depletion, water shortages, supply chain instability, higher costs. Some are already feeling the pain. Climate-fueled droughts led to failed red jalapeño chili pepper crops in Mexico, where iconic sriracha maker Huy Fong Foods of California sources that main ingredient. The crop failures caused production halts and retail shortages.[26] Across the nation in Georgia, a spring heat wave killed 90 percent of the state's peach crop; food makers either had to pay more for California peaches or stop making peach products.[27] Some CEOs are acting as rational capitalists and determining that industrial agriculture threatens their bottom line and shareholder returns. Others balance that thinking with their environmental and social governance (ESG) goals and greenhouse gas reduction pledges. At the time of this writing, fifty-eight of the world's one hundred leading food companies have either made regenerative agriculture commitments or announced pilot programs or intentions, according to Tina's research.

But if customers do not buy into regenerative products over the long term, which Tina believes people will do *only* if they realize the health advantages, then regenerative agriculture investments may not pay off. That would leave our food supply in danger. "If you start taking seriously the fact that AI just predicted that we're going to blow past 1.5 degrees Celsius [of warming] by 2030, which is a much sooner timeframe than what anybody anticipated, and all of the disruption that's going to come to global food availability, more climate strikes, the next generation not being able to have anything approaching a normal life like what you and I have had in our society, that's where I'm at," Tina says. "That's how dire and serious I'm taking all of it."

A major part of Tina's work as a Nutrient Density Alliance senior fellow is helping companies and other food system influencers understand how to emphasize the regenerative agriculture–nutrition correlation, with the goal of heightening consumer demand. "Currently, they're all focused on procurement risk, stakeholder management, greenhouse gas reporting, et cetera, and not on acting like consumer products companies. They have left the consumer out of this movement," Tina says. "And I actually think we're at a critical point with the regenerative movement where if the consumer demand doesn't come on board very strongly in a few more years, we will start to get corporate flavor of the month."

Nutrient density is not just a marketing tool, though. It is a means for holding Big Food accountable for regenerative claims. Nutrient density is quantifiable and thus an excellent way to verify that food was regeneratively produced. Regenerative farming and grazing lead to soil health (and its attendant nutrient availability, mycorrhizal fungi, and carbon) that industrial and shallow organic simply can't replicate. Healthy soil in turn results in greater nutrient density in the crops and animal products coming from that soil.[28] "It's that molecular level that is proving why regenerative ag is so crucial," Tina says. "You have biodiversity above- and belowground. You have the fact that we know that a teaspoon of healthy soil contains more microbes than the number of humans on the planet. Then you can show the mycorrhizal fungal network and its relationship to the plant at the molecular level: the uptake of the nutrients, the interplay of the carbon sequestration, which the plant then exchanges for nutrients with the soil. The nutrient-dense outcome of the food that contains more polyphenols, more antioxidants, and the right omega-3 to omega-6 fatty acid ratio that doesn't lead to chronically ill outcomes for us in dairy and animal products." Shallow regenerative farming cannot hide from precision nutrient density tests, especially if government regulations ensure that nutrient targets are high and that food makers can't fortify products as a workaround.

Nutrient density is a great way to verify the use of regenerative practices—and perhaps the most relevant for consumers—but not the only way. Soil health tests are another method, as soil and ecosystem scientist Dr. Liz Haney told me. Liz and her husband, fellow soil scientist Dr. Rick Haney, co-developed the Haney test, a first-of-its-kind assessment of nutrient availability and microbial activity to gauge soil biological health. Today Liz co-runs Soil Regen, which offers on-farm regenerative agriculture consulting, farm planning, farmer-focused educational events and trainings, and soil testing and interpretation. Soil Regen also provides a science-based regenerative certification label for farms and products. "My business partner and I do research and design for product efficacy, for transitioning to regenerative agriculture versus conventional, and use the Haney test quite a bit in helping producers improve their profitability," Liz explained. "Our goal for [the certification program] is, we believe that farmers that are doing regenerative practices should be paid a premium because they're the innovators. They're the ones out there managing all this and improving the environment. We're really trying to connect the farmer all the way through the soil health, through production practices, to the consumer, for the betterment of the environment."

Unless the federal government or another regulatory authority defines regenerative agriculture and imposes penalties for violating standards, any farmer can claim to be regenerative, just like any company can put regenerative claims on products with or without third-party verification. A complicating factor is that regenerative resists black-and-white delineations because it means adapting agricultural practices to distinct environments, executed by individual farmers with unique goals, financial means, and skills, operating within specific communities and cultures. A common critique of regenerative agriculture is that it lacks an absolute definition and as a result is ripe for greenwashing. Tina believes regenerative agriculture will gradually become more defined and that people within the movement are heading in the same general direction in terms of practices and philosophy, with some farther along on the spectrum than others.

"This is how a movement gets started," she says. "We need all these voices, everybody from the bare minimum, four or five practices that Chico State outlined in the regenerative definition in early 2017—cover crops, crop rotation and diversity, low- and no-till, living roots, animal or manure integration[29]—all the way to Regenerative Organic Alliance's organic plus–plus definition.[30] Because we need those north stars. We need to know what we should aspire to. I need people out in front proving what the system is actually able to do when you're fully in that mindset. And we need people who get onto that minimum-practice adoption end of the spectrum because their mindset gets changed when they're there. Because they're no longer doing chemically intensive agriculture, more of that profit is going in their own pocket. They see their land come back to life, they start to see the natural cycles that they've been suppressing with the chemically intensive agriculture in a way that changes their relationship to the land."

Coincidentally, my conversation with Tina in Anaheim happens on International Women's Day, a global celebration of female achievement. The day also shines light on gender inequality, gender-based violence, and the need to expand women's rights. It's not lost on me or Tina that America, on one hand, has empowered women like her to take on corporate leadership, and on the other is systematically revoking women's liberties and rallying around unabashedly sexist male leaders. With all that's been stolen from women and girls, I'm having trouble celebrating this year. Protesting, calling my representatives, moving to a blue state, or having a good old cry session all seem more appropriate. But when I hear about the particular set of gender-related challenges that Tina overcame in her career, I feel renewed hope. Because Tina shouldn't be at this conference educating others about nutrient density and regenerative agriculture. She shouldn't have climbed the ranks at Kellogg's or earned a college degree. This version of Tina, a successful woman who sits on multiple boards and testifies in front of Congress, shouldn't exist—and yet she does.

Tina grew up about an hour's drive from Battle Creek, Michigan, where her father's side of the family grows conventional row crops and dairy cattle. Tina's father worked on that large-scale operation but resisted his birthright of taking over the family farm. Instead, Tina's parents lived on an adjacent five-acre farm where the family raised livestock and grew their own produce, sourcing other food mostly at a pre-order co-op. While Tina's parents were in some ways progressive—they ate organic food before it was widely available, opted out of industrial agriculture, and built their own earth-sheltered home—her extended family and the surrounding community were highly conservative. Tina describes the area as religious and fundamentalist, with lots of Amish neighbors, a religious group that values the role women play in the household but that also demands female submission and stops their education at grade eight. Even though Tina's family is not Amish, delineated gender roles dictated the area's social life. She was forbidden to wear pants until age fourteen, she married at nineteen, and almost no one expected her to attend college or build a career. Still, Tina craved more, in part because of her mother. "I would say looking back the single biggest factor was the fact that my mom was educated, pursued her education, became a postmaster and retired as such," she says.

The local town offered little in terms of fulfilling employment, so Tina took the bold step of working in Battle Creek instead. For about five years she worked at a small staffing agency that serviced Kellogg's, the area's major employer. Seeing how many employees Kellogg's required and the wide variety of jobs there, Tina figured the company offered more upward mobility than the staffing agency, so she took a job at the W.K. Kellogg Institute for Food and Nutrition Research. She helped food scientists analyze and source ingredients, which helped her see how the food chain works and the process by which food companies formulate products around nutrition, supply, and cost. "At the time, all the global innovation for Kellogg's was happening in that building," Tina says. "So I got to work with food scientists from every continent except Antarctica for five years. I saw why you go to college. I had never before seen that in my career, ever. The scientists liked me because I was smart and I was service-oriented, so I would help them think about stuff and I would give them solutions they didn't even know they needed and by the end, they would actually be like, 'Hey, tell me which raisin or which cranberry to use because I know you know better than me.' Or, 'Hey, I gotta make my margin work on this and I'm missing out on something, show me what you got.'"

Tina's expertise and service-driven ethos helped her build strong relationships with colleagues. When a role opened at Kellogg's Kashi brand, one of the female food scientists lobbied for Tina to get it. Over and over, she says, women have made positive differences in her career. "I am very fortunate that many men, but *a lot* of women, have made a point of pulling me forward where I need to be," she says. Now she pays that support back by creating opportunities for other women, especially BIPOC and LGBTQ+ women. When Tina is asked to serve on a board, for instance, she suggests the organization consider a diverse peer alongside her. "I view it as my job—again, service—to pull, push, invite, collaborate, bring forward the things that need to be in the system. It is my duty to ensure that this movement is as diverse, represented, deepened, further extended, grown, all the words as far as possible," she says. "I'm heartened by how the regenerative movement is moving BIPOC and women, especially Indigenous communities, more to the center."

Tina is just one example of how women are not simply inserting themselves into a male-oriented food industry while leaving the structure of that industry unchanged. These women intend to build a new, sustainable model, shaped not by women alone, but by everyone who embraces holistic thinking. "In this movement, we are consciously seeing women step into a leadership space in agriculture that didn't previously exist in the same way until now. When you stepped into that space before, you may have just been stepping into the man's world of what typically would happen in a commoditized agricultural system," Tina says. "Now we're stepping into it and looking at it through the lens of inclusion, bringing systems together, impacting policy, acting at the grassroots level. I see women at all of those tables, and I continue to see women leading imperative programs at really large food companies that represent the brands that most consumers have in their fridge or pantry at any given time."

Family, community, inclusivity, equity, the environment, farmer and farmworker well-being: these are top concerns for many of the women building the regenerative food system. So is human health—a personal issue for Tina. Her grandfather died of leukemia caused by herbicide exposure. A number of her relatives suffer from cancer, autoimmune diseases, and birth defects. The family's health story, like that of most Michigan residents, is likely tied to a contamination that started in 1973, when a chemical company shipped polybrominated biphenyls (PBBs), used in flame retardants at the time, instead of a livestock feed additive to a Michigan Farm Bureau Services feed mill. The PBBs ended up on farms and retail outlets across the state and made their way into cattle, sheep, chicken, and pig feed. Meat, eggs, and dairy products became contaminated, and almost 2 million animals died.

What's worse, farmers and consumers, including unborn children exposed to PBBs in utero and babies drinking breastmilk and formula, suffered devastating health consequences. Thyroid disease, miscarriages, breast cancer in females, urinary and genital problems in males, and low Apgar scores (a measure of how well babies tolerate birth) are a few examples. As recently as 2017, 60 percent of a cohort of exposed individuals had PBB levels higher than the national ninety-fifth percentile.[31] This mass poisoning, one of the nation's largest, directly impacted Tina's family, who handled livestock feed at their Michigan dairy and ate contaminated food. Her father developed Lewy body disease, a dementia his doctors

and rot. The GMO potato's creator, Caius Rommens, has publicly denounced his invention because of these and other health and environmental concerns.[35] GMO or not, farmers spray chemicals like diquat, paraquat, glufosinate-ammonium, and others on potato plants to kill the vines prior to harvest, a process called haulm killing.[36] Scientists know that potatoes, carrots, beets, and other root crops absorb agrochemicals via leaves and soil and then store, or translocate, those chemicals in their roots—roots that humans eventually eat.

Virtually all conventional potatoes then receive an anti-sprouting fog, usually isopropyl N-(3-chlorophenyl) carbamate (CIPC), on their skin. The potato industry claims anti-sprouting residues disappear with washing and cooking, but as researchers note, "[CIPC's] continuous use is actively being discouraged because of safety concerns. It has been shown to be detrimental to both the environment and consumer health. For instance, the degradation products of CIPC, such as aniline-based derivatives, e.g., 3-chloroaniline, have been reported to be pollutants that are highly carcinogenic and toxic to the environment. These toxicological and other concerns have led different countries, notably the European Union, to progressively regulate and, in some cases, completely prohibit the use of CIPC."[37]

Many potatoes end up at restaurants and food processors that fry French fries in oil made from GMO soybeans, corn, or canola.[38] High-temperature frying produces a chemical called acrylamide, which the U.S. Environmental Protection Agency classifies as "likely to be carcinogenic to humans." Fast-food restaurants then serve those fries in containers that could contain per- and polyfluoroalkyl substances, forever chemicals with dangerous health effects.[39] "And yet we think of French fries as just being potatoes and oil and salt," Tina says.

By the way, potatoes are not the only conventional crops desiccated before harvest. Some farmers spray wheat, oats, barley, lentils, peas, non-GMO soybeans and corn, flax, rye, triticale, buckwheat, millet, canola, sugar beets, and sunflowers with glyphosate, in some cases to induce uniform and quicker ripening, and in others to clear plant leaves and other debris to make harvest easier.[40] And this is *after* most farmers spray fields with glyphosate pre-planting (post-planting for GMOs) to control weeds. Glyphosate shows up in food made from these crops, like breakfast cereal, pizza, granola, crackers, bean products, and pasta.[41] This is troublesome because the chemical is associated with an elevated risk for non-Hodgkin lymphoma in humans and potentially other cancers and health problems.[42]

Glyphosate also negatively impacts grain quality. One major mill, Grain Millers, Inc., linked performance and texture problems and less beta glucan (a fiber that promotes lower cholesterol and better cardiovascular health) in oats that had undergone glyphosate desiccation. The company now refuses to buy desiccated oats.[43] "2,4-D, glyphosate, dicamba, all three of those are actively used in desiccation. They enter our food system without any notification that those things are there," Tina says. "We tend to think of those three things being sprayed at some point early in the crop production cycle and burned off or metabolized, or the soil has worked it out by then. And yet, it may be actively sprayed right before that crop becomes food."

Desiccation is emblematic of a food system where profit is king. Where food makers and the government bend to the will of corporations who want to keep selling agrochemicals and GMOs that science tells us are unsafe for people and the environment. Where farmers are so brainwashed by Big Ag that they will poison consumers, vote against their and their community's interests, degrade their soils, and risk their family's health. Where the USDA essentially commands those same farmers to either get out or go into debt so they can be industrial. Where Big Food took over the organic industry and made it into a less-bad version of conventional agriculture. Where even regenerative brands concentrate on moneymaking UPFs rather than minimally processed products. In our capitalist system, trusting global food and agriculture companies to adopt an authentic version of regenerative agriculture is hard for me. Yet the alternative, a world in which Big Food and Big Ag maintain the status quo, is frightening.

The uncomfortable reality is that large food companies manufacture the majority of our food. For most consumers right now, opting out of the Big Food world entirely is neither possible nor convenient, and for those who are food insecure or living in a food desert, opting out is not affordable. Big Food has acquired so many organic and sustainability-focused brands that even when we think we are opting out, we might not be. The same company that owns Wonder Bread owns Dave's Killer Bread. General Mills owns Cascadian Farm. Campbell Soup owns Plum Organics and Pacific Foods. Hormel owns Applegate Farms.

Clorox owns Burt's Bees. Breaking up Big Food and increasing diversity and localness in the food system will take time—and we are running out of time when it comes to reducing carbon emissions, preparing our food system for climate change, and addressing the food-related public health crises we are experiencing.

One-third of global greenhouse gas emissions come from the world's food systems. Of that one-third, growing crops and livestock account for 71 percent of emissions, and the other 29 percent come from packaging, transporting, retailing, and disposing of food.[44] Big Food's on- and off-farm regenerative commitments should have happened decades ago. Still, the chance to reduce one-third of total world greenhouse gas emissions represents a massive opportunity within just one industry that a relatively small number of major players control. Large companies have the power to usher in sweeping change with their money, market leverage, and social currency. If Big Food demands regenerative production, then the food system will respond at scale, just as it did when those same companies demanded cheapness and uniformity.

"Regenerative ag *has* to scale in order for it to work," Tina observes. "If it's just 1 percent of acreage in the U.S. like organic, or 10 percent of the food system sales like organic, or any other food sector that you can slice off, then it's not going to have the type of impact that we really, fully need at scale. You should want it to scale to all corn, soy, wheat standard rotations in America, the Chicago Board of Trade commodities. You should want all of that to pivot to a continuous living cover, no-till, animal integration where possible, bioactive, nature cycle restored food system because you've restored the water cycle, you've reduced the volatility in your local climate and your food system, your risk, the insurance that we have the farmers carry. You've positively influenced all of those metrics if you've installed a regenerative system," she says.

Just like scaling regenerative agriculture requires a wide swath of farmers, it also needs the broadest possible coalition of food and agriculture companies. Scaling requires the Walmarts, Nestlés, and Cargills of the world all the way down to your neighborhood bakery, and everyone in between. Yes, consumers, shareholders, and third-party watchdogs should hold companies accountable, Tina says. But she also believes authentic change is taking place in Big Food, in part thanks to new, more diverse leadership. She encourages skeptics not to give up on the food industry. "I will have people say to me, young people, 'Oh, you worked at a big food company, I could never work at a big company, they're all evil.' I'm like, 'Well, wait a minute. Are you telling me that you're only willing to let people who don't have human-centric values go work there and you're not willing to also bring your value set into that large company to try and be part of the balance for what that company is capable of doing?' Taking that 'big is bad' attitude into the space when we actually need more people than ever to undergo this transformation, I just don't have a lot of patience for that anymore," Tina says. "My take and my hope, and the reason that I continue to work in a food system this large, is because if you can make changes at this level, you will have really, truly impacted something."

Tina sees several concrete ways capitalistic principles can advance regenerative agriculture. Companies could tie compensation to regenerative outcomes, for example, and elevate positions like chief sustainability officer to the same prominence as chief financial officer, with similar expectations for results. "Until C-suite compensation is impacted by something, it does not get done. I don't care what the company calls itself from a values perspective. It does not get done," she says. Companies can leverage the superior quality of regenerative ingredients, like longer shelf life and higher nutrient density, to improve their bottom line by promoting these attributes to both retailers and consumers.

A food company's profits are usually underpinned by a stable supply chain—wild price and availability swings eat into returns—so companies have a reason to want the reliability that regenerative production offers in a changing climate.

And lastly, regenerative agriculture can support emissions reductions targets. At the time of this writing, 3,383 companies and financial institutions have adopted plans for lowering emissions through the Science-Based Targets initiative (SBTi), according to its website. Of those, 2,371 made net-zero commitments.[45] Almost any company involved with agriculture and food can reduce its emissions by removing industrial agriculture practices from its value chain.[46] Public pressure is on to meet the initiative's targets, pressure that will only increase as climate change accelerates and companies risk losing investors' money and facing backlash by failing to position their operations for a hotter planet. "By the end of this decade, people will be pointing fingers," Tina says. "Regen ag is almost single-handedly their—it's too simple to say get-out-of-jail-free card—their biggest lever to pull to not end up in that future."

When it comes down to it, that is what regenerative agriculture is about: the future, a better one than humankind is on track to bring about. The future I catch a glimpse of at Expo West is both promising and concerning. There is much hope for change in those food-filled rooms, as well as clear warnings about the forces of capitalism. Tina and I say our goodbyes and I head outside. Somehow, I am hungry again, so I turn in the direction of the nearest exhibit hall. Wafting from the food court exhaust pipes into the chilly spring Anaheim air is the telltale smell of fast-food French fries.

money from something, they are disinclined to stop, even if those investments threaten their or other people's livelihoods. Again, consider fossil fuels; despite the fact that carbon emissions endanger the planet's stability and humankind's survival, returns keep investors hooked. Big Oil and its profit-at-all-costs ethos are to blame for this behavior, but so are fund managers, endowment overseers, investment banks, and other institutional decision-makers who continue investing in fossil fuel-dependent sectors. Even individuals vote with their retirement account allocations, although we might forgive some of this because company 401(k)s and the like may limit investor options. Still, we can't ignore the reality that money can have a corrupting influence, as it did within some areas of the organic industry. For these reasons, touting the profit potential of regenerative agriculture (which is very real, by the way) worries me. Too often, investment enables exploitation.

But as with the notion of Big Food staying out of regenerative agriculture, the prospect of investors maintaining the status quo, directing money to industrial agriculture and its related entities, is scarier. Failing to shuttle new and existing capital into a regenerative food and agriculture system will surely leave us more vulnerable in a changing climate. I'm not the only one who feels this way. Sarah Day Levesque of Regenerative Food Systems Investment has spent the last five years catalyzing capital, connections, and conversations around investment in regenerative food systems. She believes agriculture and food systems offer unique opportunities to combine sustainable wealth creation with healing the planet and ourselves. "Watching the way we can use capital to kind of re-create the systems that are hurting us right now and build so much value out of them, that keeps me going," Sarah says. "I feel from an impact standpoint, there's no better place to be than trying to rebuild a food system because it touches so many different things, like climate, biodiversity, overall human health, and farmer well-being."

Sarah grew up in California's Sacramento Valley, an agricultural powerhouse that pumps out conventional almonds, stone fruit, rice, produce, and livestock, made possible with extensive groundwater irrigation. She attended University of California, Davis and in 2006 earned dual master's degrees in international agriculture development and agriculture and resource economics. Sarah took a job at an agriculture consulting agency analyzing intellectual property traits for agribusiness companies like Monsanto and DuPont. The agency eventually melded into a media company, and Sarah stepped into agricultural news, content development, and event planning related to agriculture investing, grain and oilseed production, women in agribusiness, and other topics. While she enjoyed her work, she felt guilty about advancing industrial rather than sustainable food and farming. "I spent about ten years total with that company working in ag commodity supply chains and agricultural investing, so that's where I got into this investing side of things, and I had that ag econ background," Sarah explains. "I really liked the idea of capital moving the food system forward, but it was time for a career change and, going back to my grad school days, I really thought I was going to be doing something more impactful. I love the food system, but it didn't quite feel like we were making a difference."

In 2016, a general manager of events and media position opened at Acres USA, a publisher dedicated to advancing production-scale organic and regenerative farming. The company also hosts an annual Eco-Ag Conference and Trade Show and dozens of other educational events. Sarah jumped at the chance to lead the company's events and media division, and she soon found herself at the center of the conversation about how to expand regenerative agriculture. Two lessons became clear. One, regenerative agriculture works—farmers, researchers, economists, and others prove this repeatedly—and it is scalable. Two, significant barriers to entry stand in the way. Farmers require monetary support, educational and technical resources, cultural shifts in their communities, and assured markets. But resources are not easily flowing to stakeholders addressing those barriers, or to enough farmers. The issue isn't whether regenerative is feasible, but whether the food and finance system can catch up.

To Sarah, an agricultural investing and economics expert, the missing piece is capital. "If we want regenerative to grow, then we have to create mechanisms for the capital to flow to the farmer, to flow to the food companies, to flow to the tech companies, all those people and stakeholders," she says. "If we want to be truly regenerative, you have to have the whole ecosystem in place. You can't just have one part, one place along the supply chain to do that." Acres USA's parent company agreed and tasked Sarah with finding solutions. The result was Regenerative Food Systems Investment, an organization that works across diverse groups of funders and stakeholders to mobilize capital for regenerative food and agriculture projects. The organization also aims to expand the distribution of information and tools that communicate the financial, ecological, and social benefits of investments in a regenerative food system.

Sarah launched RFSI in 2018 and serves as its managing director, while continuing to function as an adviser for Acres USA. RFSI is not a registered investment, legal, or tax adviser or broker/dealer (an important distinction from a Securities and Exchange Commission perspective). Instead, Sarah and her team are matchmakers of sorts. They provide opportunities to unite the people constructing a regenerative food system with the people seeking to accelerate that construction with capital. RFSI does this primarily through in-person and virtual events. The marquee gathering is the annual RFSI Forum, which brings together investors, fund managers, foundations, entrepreneurs, food companies, agriculture and food service providers, farmers, advocates, and other stakeholders. The two-day assembly attracts nearly five hundred attendees and features informational speakers, round table and panel discussions, pitch sessions, case study presentations, and networking opportunities intended to link capital with regenerative projects.

Rather than a "preaching to the choir" event where regenerative agriculture missionaries talk to existing believers, the forum is intended for novices, people with regenerative agriculture experience, and anyone in between. "Our goal really is to expand the community of funders and capital activators. We want our community to be in there, but we really want to reach outside of our community and say, 'Who are those actors that are curious about the regen space and need to learn about the capital equation and how that drives overall development of the space?' It really is different than another investing event in the sense that it's not just, 'Apply the same investment principles to different kinds of assets or projects,'" Sarah says. "It's really, 'How are we thinking about it differently and how does that influence approach?'"

This year farmers, ranchers, and people representing brands or early-stage startups can attend RFSI Bootcamp, a one-day pre-forum course to help newcomers better understand and engage in the regenerative food system investment landscape. RFSI Bootcamp grew out of a disconnect Sarah observed between farmers, regenerative agriculture innovators, and their potential financers. Farmers and innovators do not always understand investor needs, how fund managers interpret risk, or how different capital tools work. At the same time, investors may not appreciate farmer or entrepreneur experience and the barriers for entry into regenerative agriculture, or how regenerative agriculture differs from other types of investments. RFSI Bootcamp's 101-level training is intended to bridge those gaps.

"Farmers might be really aware of the EQIP [Environmental Quality Incentives Program] loans and other funding that they can get from the USDA," Sarah explains. "But what if you need to capitalize either your farm or your company, and you've never had to accept outside or non-USDA capital? We want to help expand understanding of what the entire regen capital ecosystem looks like. What each kind of player is and what they look for, so you can better understand who you should go after. And then here's how you think about how each kind of capital fits within your operation. That's really trying to bring those stakeholders closer to the funder or investor. Then on the other side, we're going to have the same kind of 101- and 201-level training at the main event, which will dig into how we're investing in regeneration today. This is what it looks like versus if you are investing in venture capital outside of agriculture. And then obviously curating more of those conversations will be part of that as well."

The farmer-financer knowledge gap is understandable given the complexity behind the word "capital." Put simply, different types of capital can fund regenerative agriculture and the food system. The main capital categories (outside of government and public capital) are debt, equity, and philanthropy. Debt is exactly what it sounds like: borrowed funds to be repaid later. Equity means investments in real assets, such as farmland or infrastructure, or in businesses, such as startups. Philanthropy typically means grants intended as gifts or given in exchange for lower-than-market-rate returns. Individual and institutional investors use many different entities or vehicles to deploy their capital. For example, they might put money into private equity funds or private credit funds, or distribute grants through donor-advised funds.

The growing realization among altruistic investors that regenerative agriculture addresses an incredible range of environmental and social issues they care about is a major force pushing capital into regenerative agriculture, Sarah tells me. In the past, investors might have directed their money toward individual concerns, like habitat or wildlife conservation, rural economic growth, or carbon reduction. But a growing number of people are using regenerative agriculture investments to check the box, so to speak, on numerous issues at once. Many investors hope to generate what the Croatan Institute calls "soil wealth," a term for the collective riches of regenerative agriculture.

Soil wealth is, of course, healthy and productive soil, but it also includes enhancements in biodiversity, water quality, farm and landscape resiliency, and carbon sequestration. Soil wealth means decreased greenhouse gas emissions, higher household incomes, lower unemployment rates, better human health outcomes, increased social and racial equity, humane animal treatment, a more stable food supply, and a narrower urban-rural divide.[1] "What we are increasingly seeing people understand—which is great—is that if you invest in regenerative food systems, you are also investing in climate, biodiversity, human health, and farmer well-being. So you can amplify your impact," Sarah says.

What does all this capital look like for regenerative farmers and businesses across the supply chain? The answer varies. It might be low-interest loans, flexible-payment loans, cash grants, or loan guarantees to fund a farmer's transition to or expansion in regenerative agriculture. It might take the form of "integrated capital," which is the coordinated use of different types of financial and nonfinancial transitions."[2] For instance, industrial production depends heavily on finite inputs like irrigated water and fossil fuel; as nonrenewable resources grow more expensive and eventually disappear, investor returns likely will shrink. Soil loss is another risk associated with industrial production. The U.S. soil erosion rate is as high as ten times the natural replenishment rate, largely due to industrial practices.[3] Worldwide, about one-third of soil is degraded, and topsoil in many regions could be completely gone by 2074 if current degradation rates continue.[4] Farmers compensate for erosion and poor soil health with synthetic fertilizers, another fossil fuel–derived input, but those won't be an option much longer as oil supplies dwindle. Regenerative agriculture, on the other hand, restores soil health, conserves water, and requires far less fossil fuel and other inputs, which all helps reduce risk and increase long-term profitability.

Input dependence also leaves farmers vulnerable to high prices and scarcities that cut into profit. Take the fertilizer situation after Russia invaded Ukraine in 2022. That year the war and lingering effects of the COVID-19 pandemic prompted natural gas prices to reach their highest levels since 1973's oil crisis. Because natural gas represents 75 to 90 percent of production costs for fertilizers like urea and ammonium, natural gas price jumps quickly led to fertilizer shortages and cost increases for farmers as manufacturers slowed or stopped production. Additionally, Russia accounts for 19 percent of global potassium exports, 14 percent of phosphorus, and 16 percent of nitrogen, the three main ingredients for conventional fertilizer. War-related sanctions on Russia further contributed to fertilizer price spikes and supply shortfalls.[5]

Geopolitical conflict is likely to occur more frequently in a warmer world, which could make input shortages and cost hikes like this more common. In contrast, regenerative farmers rely on fewer inputs sourced from home and abroad. This self-sufficiency not only lowers their financial exposure, but also reduces the nation's risk of a food system disruption or collapse due to conflict. Compared to the current system, a regenerative system would rely on many more local and regional farmers and food makers, which would further shield the U.S. food supply from events outside the nation's borders. From this perspective, our industrial, consolidated food system is a national security problem that regenerative agriculture can help solve. A nation that can feed itself is strong and resilient—and we need those traits more than ever to withstand warming-induced geopolitical challenges on the horizon.

But the biggest risks for industrial agriculture are the on-the-ground effects of climate change. "Climates differ and plants vary, but the basic rule of thumb for staple cereal crops grown at optimal temperatures is that for every degree of warming, yields decline by 10 percent," writes David Wallace-Wells in *The Uninhabitable Earth: Life After Warming*. Warming-driven insect pressure, funguses, and diseases will cut yields as well.[6] Because conventional farms tend to overspecialize in a small number of crops or livestock on a large scale, pivoting to something different in response to changing conditions is financially and physically burdensome. Those farms also have few other revenue streams to compensate for income streams that fail. All this translates into profit loss up and down the conventional food chain.

These agriculture-related losses could happen concurrently with financial hits across economic sectors if other industries fail to adapt to global warming, too. According to the Brookings Institution, 215 of the world's largest companies together face up to $1 trillion in climate risk that is both transitional (shifts in innovation, tech, regulations, competition) and physical (wildfires, floods, sea-level rise, hurricanes).[7] The SEC under the Biden administration wants publicly traded companies to disclose their financial risks related to climate change like they do other risks. Should climate risk disclosure become law, the food and farming industry will have to reckon with industrial agriculture's exposure, as will investors. Transitioning the food system to regenerative now, rather than waiting until farms become too unproductive and unprofitable to stay in business, is clearly the best option, and not just from an investment standpoint.

Another challenge is supporting farmers more effectively with capital. Sarah sees increasing investment focus on farmer conversions to regenerative—a great development, but she believes farmers need additional and more suitable funding vehicles, and way more supply chain investment. "There is more funding going to venture capital and folks who are working on technologies to advance, say, inputs like biologicals, as a tool to help farmers transition," she says. "But there's less attention—and this is where we wave our flag at RFSI—going to what happens once you get a farmer to transition. Where is that product going? Where's the infrastructure? And what brand or CPG [consumer packaged goods] are the ingredients ultimately going to?"

Lack of investor focus on the supply chain, and too much focus on developing new inputs and tech, is leading to two parallel food systems in the regenerative world, Sarah explains. In the Big Food system, corporations are working directly with large farmers, assisting or just encouraging them to convert to regenerative and promising to buy their outputs. In the non–Big Food system, small and midsize regenerative farms that do not work with corporates can't find buyers or access resources to transition. And on the opposite end of the chain, emerging, small, and midsize brands can't find regenerative ingredients.

Other market gaps are hindering the regenerative movement, too. For instance, few outlets exist for harvestable cover crops if farmers cannot "sell" them through livestock that eat them. Cover crops lead to outcomes that enhance a farmer's bottom line, like soil fertility, water infiltration, and natural pest management. But the market is missing opportunities to incorporate certain cover crops into the food system and financially reward farmers for growing them. The system also needs more buyers for lesser-utilized but drought-tolerant crops like millet, amaranth, Kernza, and buckwheat. Similarly, farmers who convert cropland back to native grassland ought to be rewarded with opportunities to market grass-based products like native plant seed, hay, and livestock.

Using capital to design shorter and more diverse supply chains will be a top-line subject at the next forum, as it continues to be in RFSI content and resources. "We call it internally the unsexy middle, the middle of the supply chain—it's the plumbing of the food system and it's not super attractive to investors. There is a lot of risk and complexity associated with the middle, so it's really hard," Sarah acknowledges. "But there's an increasing group of people, in our community at least, who are really looking at saying, 'Let's solve it.'"

I asked Sarah for an example of someone who is saying "Let's solve it." She pointed me to Esther Park, CEO of Cienega Capital. Cienega is a family office with the mission of investing to improve soil health, regenerative agricultural practices, and local food systems. A family office is a private wealth management firm created to provide financial and investment services for an ultra-high-net-worth individual or family, or a small group of such families. Cienega Capital's founder is impact investor Sallie Calhoun, formerly of Globetrotter Software, a tech company she and her husband founded and ran for decades. Cienega Capital is part of the #NoRegrets Initiative, a regenerative asset strategy Sallie established that also includes the family's Globetrotter Foundation (philanthropy), the Paicines Ranch (regenerative livestock and land management on 7,600 acres), and the Paicines Ranch Learning Center (nonprofit).

The mission of #NoRegrets is to move social, ecological, and financial capital toward soil health to address climate change. Esther's job is to fulfill that goal through the investment arm, Cienega Capital. "We invest across the supply chain with a real bias towards farmers. About 30 percent of our portfolio is farmers, to farmers directly," Esther explains. "Anybody else in the supply chain that we support has to have some kind of really direct tie to the farmers, or at least a really good aspiration towards that. Companies that are just anonymously sourcing organic ingredients, that's not really interesting to us. But, say, Lotus Foods, who works very directly with their farmers, provides technical assistance, helps them implement these conservation practices, that is what we want to be doing."

Lotus Foods imports heirloom rice that they either package and sell, or use to create rice products such as noodles and soup cups. The company sources from smallholder farmers, pays organic and fair-trade premiums, and advocates for a more equitable food system. They are especially concerned with using regenerative practices to reduce the amount of water farmers need to grow rice, an urgent matter in a changing climate. These traits, plus the close farmer connections Esther mentioned, convinced her to invest in Lotus Foods.

The investment is paying off in multiple ways. A good portion of Lotus Foods' rice imports come from China, and Trump-era trade-war tariffs and pandemic supply chain snarls took a toll on their business. Lotus Foods began looking for domestic rice suppliers in response. "In the process—and I wouldn't say that we're taking credit for any of this—but in the process they have teamed up now with a not-for-profit organization who's also a grantee of ours called Jubilee Justice. What they are doing is, they're creating the first System of Rice Intensification farmers in the United States among a cohort of Black farmers in the South," Esther says. The System of Rice Intensification is a holistic approach to growing rice that minimizes water use and alternates wet and dry conditions to boost yields and reduce methane and carbon emissions.[8]

"Lotus Foods has exploded since we made that initial investment, and I would say they are seven times the revenue that they were when we first started," Esther says. "Again, I don't take any credit for that. But it's amazing to see a company do well like that. We did provide them with a critical piece of capital when they needed it. They were able to grow quite rapidly over the last few years. They were able to survive all the supply chain disruptions, not just the tariffs but all of the port delays, the labor strikes, all of that, and are now working with this nonprofit organization to really develop some historically marginalized farmers into their supply chain."

Lotus Foods illustrates how Cienega is helping build out a regenerative supply chain. But the main focus, as Esther mentioned, is direct farmer support. In some cases, the products are traditional, like term loans, working capital, convertible debt, or equity investments. In other cases, Esther tailors capital to match farmer needs, perhaps by adjusting the terms or bundling several types of funding. This is an example of the flexible finance options regenerative producers need from institutions, too. Cienega's largest project revenue-wise is $50 million, but the majority are smaller. For example, in one case Cienega bought farmland, leased it to a regenerative farmer, and arranged for him to purchase that land. In another instance, the firm participated in a joint venture with a rancher. "Those are very customized situations where it was like, financial, ecological, and social balance. Others may fall short financially but achieve incredible ecological and social outcomes. "This approach muddles returns, and some people will say it's not an institutional investment approach and will never scale. But that's the idea," Esther continued in the editorial. "We don't need to scale some huge investment vehicle—we need a thousand people (or more!) to invest this way." This method of investing, and the call for more people to do it, is at the heart of the #NoRegrets Initiative.[10]

Esther's philosophy and the #NoRegrets strategy as a whole might also be called holistic investing, responsible investing, values-based or ethical investing, or impact investing. Others refer to it as the triple-bottom-line approach that takes people, planet, and profit into account. Whatever one calls it, women are driving this style of investment. Many women investors (who increasingly have more investable assets of their own and decision-making power over shared assets) are putting money into environmental and social governance (ESG) funds and other holistic investment vehicles. Female investors are twice as likely as male investors to say it is important that the companies they support have ESG outcomes. Women are also more likely to prioritize ESG impact when choosing investments, whereas men are much more likely to prioritize returns.[11]

That's in part because women tend to break down the silo between financial investments and social impact, argues Julie Davitz, former Bank of the West head of impact solutions. "One reason it might be easier for women to take this holistic view is because of their different attitude toward investing. Many studies have shown that women and men tend to have different investing styles. Men are more likely to focus on the short-term performance of individual investments, while women are more likely to keep their eye on their financial goals. Seen in this context, the need for investments to make a positive impact is just one more goal to consider," Davitz writes.[12]

It's not that women do not care about or generate returns. The Warwick Business School in the United Kingdom found that female investors outperform men by 1.8 percent,[13] and U.S.-based Fidelity Investments determined that women outperform men by 0.4 percent.[14] Women do this by steering away from speculative stocks, holding their investments longer, and favoring diversified funds rather than individual stocks. So women seek returns but tend to understand returns as multidimensional rather than one-dimensional. Esther confirms these findings with her real-world observations. "It's easy to generalize, but I would say that for the most part in my interactions, it's been the men that have been the hardest to convince out of linear, one-dimensional thinking. It's not entirely their fault. We have a society that really prioritizes that kind of thinking, so it's just an easy rut to fall into. I still fall into that rut. I think we all do to some extent. It's the culture that we've grown up in," Esther says. "But I think that women naturally are able to think more broadly and embrace complexity in a way that's very different from dominant male thinking around linearity."

In addition to being individual investors, women hold more leadership and decision-making positions within ESG investing compared to other sectors of the finance world, often directing entire ESG units for firms.[15] ESG standards assess a publicly traded company's business practices and performance on environmental and social issues. Independent ESG research firms produce ESG scores for public companies so investors can align capital with their values. Investing in companies with high ESG scores or ESG-specific funds is one of the only ways everyday people without significant wealth can participate in sustainable ventures—and people are increasingly open to ESG participation. That trend exploded when BlackRock, the world's biggest asset manager, committed to ESG investing back in 2020. If public companies want the capital that firms like BlackRock provide, then meeting ESG targets is a must.

Journalist Alastair Marsh writes in *Forbes* that ESG investing is "the rare area in finance where the gender imbalance in top jobs isn't hugely lopsided in men's favor."[16] And it seems these women investors are good at what they do. Forty-six percent of traditional funds available ten years ago still exist, while 77 percent of ESG funds still do.[17] Not only are ESG funds performing well, but they are also increasing as a percentage of total investment dollars. The United States had $67 trillion in assets under management in 2021. The nation's ESG assets under management is projected to almost double from $4.5 trillion in 2021 to $10.5 trillion in 2026—which means women are likely to oversee a bigger slice of the U.S. investment pie than ever before.[18] This doubling also puts women at the forefront of the regenerative transition, since regenerative agriculture and ESG investing go hand in hand.

That women ended up in ESG leadership has several explanations. One is the female investor traits mentioned earlier: holistic perspectives, long-term thinking, and a desire for a blend of social and financial returns. ESG investing requires all those qualities. Another explanation is that, until recently, ESG investing was seen as less-than, a niche sector geared toward a small number of investors that generated lackluster returns compared to "real" strategies. Many viewed ESG as emotionally influenced investing, and men tended to shy away from it as a result. Firms may have put women in ESG roles because those positions were understood as inferior or feminine—which could not be further from the truth. ESG roles were and continue to be challenging because they require innovative strategies and nontraditional thinking.

As Bonnie Wongtrakool, a portfolio manager and global head of ESG investments at Western Asset Management in Pasadena, California, put it, "ESG was not always popular or glamorous. If you have an important but thankless task, you give it to a female, because you know they will get it done."[19] Now that ESG investing is mainstream, women are enjoying newfound power and influence in the financial world. The fact that women lead the ESG sector might partially explain why some conservatives (usually male) attack ESG, since it is a source of economic and social progress for women.

Within regenerative agriculture investing specifically, Esther is seeing more women in leadership positions, but not enough. "We are increasing in numbers and it's helped us to find each other, which has been great. But the number is still so small relative to the entire population," she says. Sarah's analysis is similar when it comes to the regenerative agriculture and finance spheres. "I was looking at the stats and I was like, 'You know what, there is actually female representation in this sector.' And I think the regenerative space and the food space have much higher representation of women than other sectors. But it's when you dig in a little bit deeper to who's getting the leadership roles, and does that seem to be equitable, are they given the same chances, that's where you see the discrepancy still."

Similarly, women of color are becoming more present but are still underrepresented, as they are in the broader financial landscape. As an Asian American woman, Esther is often the only woman of color, and sometimes the only woman, in the speaker lineups at regenerative agriculture investment conferences or events. Sometimes she is one of just a few women or women of color in the entire room. While Esther has encountered intimidating environments, as a seasoned leader she is not afraid to address disparities and challenge the investment world to diversify. "Sometimes I still have to say the hard things. There's that element of feeling the burden of doing that as one of the few women and one of the few people of color onstage to do that. So there's the responsibility and the burden of it. I get kind of excited by doing it now," she jokes.

Helping women gain Esther's level of confidence and professional success is yet another element of Sarah's work at RFSI. This year, RFSI is holding its first-ever Women Transforming Food and Finance event. The day-and-a-half gathering will bring together female leaders in agriculture, food, and finance for industry learning, professional development, and relationship building. Goals include generating solutions for a new food and finance system, increasing diverse female leadership, driving funds to female-led initiatives, and developing fruitful, lasting connections.

"We already know that women are under-represented in the financial space. They're underfunded in the regenerative space, as well as across venture capital as a whole. But we also know that research shows that women in leadership have as good or better operating results and that female investors have as good or better returns than their male counterparts," Sarah explains. "We want to bring women together and say, 'We see you.' The food system is in the midst of significant transformation and women must play a key role in leading us through it. Both our food and finance systems need to be addressed. How do we build the infrastructure, processes, policies, and collaborative relationships that will allow women to successfully lead this?"

Studies on leadership styles reveal that, compared to male leaders, women leaders tend to be more participative, democratic, and collaborative; to value positive relationships, positive reinforcement, and communication; and to prioritize the public good and egalitarianism.[20] One study looked specifically at how women advance sustainability within companies. Those researchers found that women contribute to the promotion and implementation of responsible environmental and social practices by empathizing, listening, collaborating, thinking holistically, managing complexity, and being inclusive.[21] Studies generalize, of course, and do not represent all women. This book's intent is not to essentialize or promote new stereotypes and expectations, but to acknowledge the documented differences in how women tend to lead—and to argue that this style is exactly what the regenerative food system needs from its leaders.

An elevated female presence in the regenerative agriculture world compared to the industrial is already yielding positive effects, greater collaboration being one. People across the regenerative sector—of all genders, not just women—tend to cooperate rather than compete, as Sarah notes. "When you come to regenerative, it's like we're working towards a mission together, so there's that inclusivity in general, whether you're talking about gender or just willingness to collaborate," she says. RFSI is, at its core, a collaborative endeavor led by a natural collaborator. It strikes me how different such collaboration is from "get big or get out" industrial agriculture, which pits farmers against one another and against their environment, and from winner-take-all investing, which extracts profit at any cost to benefit individual investors and firms.

Collaboration means teamwork, alliances, partnerships, relationships, cooperation—basically the opposite of America's so deeply engrained individualism. Individualism has its place and usefulness. But as the controlling philosophy of a food system, it does not work. The food system touches the entire world and every person in it. Failing to work together inevitably leaves some part of the environment or society vulnerable. Collaboration—the kind women often practice, but that we *all* are capable of—is one of the best ways to ensure that the new regenerative system we're building promotes health and balance from the ground up.

8

Crest

Vanessa García Polanco, National Young Farmers Coalition

Crest: the highest point of a wave above the still-water line.

I arrive in Washington, DC, on a sunny but windy late afternoon in mid-March. Walking the elegant, intimidating city gives me a head rush. I feel power close at hand, power enshrined in monuments, power running like electricity in the currents of well-dressed people on sidewalks and in hallways. Inside the buildings around me, lawmakers and staff assigned to the various House and Senate Appropriations Committees are negotiating, arguing, learning, and hopefully compromising. Even the mundane feels significant, knowing the decisions made here reverberate nationally.

I haven't seen DC since high school, when I traveled here for a week via the National Rural Electric Cooperative Association Electric Cooperative Youth Tour. Sixteen years old, my first time on a plane. Returning at thirty-five, I am not sure if my memories of the memorials and government buildings are actually mine or if they came from photos and videos I've viewed since. I absorb the city's magnificence and history as if for the first time and think this might be a pleasant side effect of getting older: I've forgotten places and events so thoroughly that I can redo the wonder of first experience.

I'm in DC to bear witness, if only for a day, to the building of the Farm Bill. The Farm Bill is an omnibus, multiyear funding package passed by Congress roughly every five years, and it is the main way the federal government shapes the agriculture and food industry. Lucky for me, legislators are writing a new Farm Bill as I am writing this book. Early the next morning, I head to Capitol Hill to see Vanessa García Polanco, director of government relations for the National Young Farmers Coalition. The mission of Young Farmers is to "shift power and change policy to equitably resource our new generation of working farmers." Young Farmers advocates for sustainability programs, climate action, water justice, immigration and labor rights, farmer health and well-being, and access to land, capital, and USDA resources for small and diverse young producers. Shortly before my visit, Young Farmers hosted more than one hundred producers at a fly-in event here in DC. Farmers voiced concerns and priorities that inform the organization's advocacy—a true grassroots effort.

Right now, Vanessa is laser focused on the Farm Bill, since that is the best chance for achieving producers' policy goals. The Farm Bill is actually a package of mandatory and discretionary spending bills. The legislation impacts many parts of the food and agriculture world: commodity crop aid like disaster relief and insurance, nutrition assistance, conservation, research, specialty crop initiatives, bioenergy programs, trade, forestry, rural development, and more.[1] The USDA carries out most of the Farm Bill's directives. So if the legislation appropriates money for regenerative agriculture incentives, for example, then the USDA designs programs farmers can participate in that satisfy the bill's directive. As the National Sustainable Agriculture Coalition puts it, "The farm bill sets the stage for our food and farm systems."[2]

The Farm Bill wasn't always so wide-ranging. Launched in the

Dust Bowl era, the legislation originally allocated help to farmers growing commodities like corn, cotton, wheat, soybeans, rice, dairy, peanuts, and sugar. However, the bill quickly became a vehicle for propping up industrial agriculture. Over time USDA policies forced farmers to "get big or get out," a philosophy put into those exact four words by Earl Butz, U.S. secretary of agriculture from 1971 to 1976. During the 1970s, "policy objectives shifted from supporting family farms to promoting the industrialization of agriculture. . . . The farm policies of this era were designed specifically to support, subsidize, and promote specialization, standardization, and consolidation of agricultural production into ever-larger farming operations," writes agricultural economist John Ikerd.[3]

Vegetables, grain, poultry, fruit, nuts, dairy—almost everything America grows comes from an ever-smaller number of ever-larger farms. Cow-calf producers are an exception; their operations show little consolidation (remember that cow-calf operations are not concentrated animal feeding operations or slaughterhouses, both of which are highly consolidated food system components). To be fair, families own and run most of the nation's big farms that are responsible for the bulk of production.[4] Family operations are ideal, but the problem is that the number of farm families out there has been shrinking for decades as "successful" families have consolidated their holdings and pushed out "unsuccessful" ones. Such consolidation has decimated rural communities and led family farms to operate like industrial enterprises.[5]

Not surprisingly, when the government demanded all-out commodity production, oversupply was the result and prices fell, driving more farms out of business and spurring the devastating 1980s farm crisis. The ones that survived responded by expanding, but to this day, all-out production and its attendant industrial practices are so financially ruinous that most commodity producers could not stay in business long without government support, including from the Farm Bill. Longtime food and agriculture writer Tom Philpott documents this reality in *Perilous Bounty: The Looming Collapse of American Farming and How We Can Prevent It.* His analysis of USDA Economic Research Service data reveals that between 1996 and 2017, net return per acre for corn was negative for sixteen out of twenty-one years.[6] That is why the United States subsidizes farmers through price and revenue supports, crop insurance, low-interest loans, government purchases, and more; for a while the government even offered farmers direct payments. The majority of agricultural aid goes to the Big Five: corn, soybeans, wheat, rice, and cotton, with corn subsidies way, way higher than subsidies for the other four crops. The USDA, the Department of Energy, and federal legislation also pushed farmers to grow corn and soybeans as biofuel and livestock feed, not as food edible without processing first. All these incentives drove more farmers into commodity production. Meanwhile, "specialty" crops like vegetables, fruits, nuts—outputs people can eat directly—receive fewer subsidies combined than corn does on its own.[7]

Food production justifies government support. Our nation should be able to feed itself for security reasons. Producers deserve a dignified life and citizens deserve enough to eat. The problem is not government action, but the wrong kind of action. We ought to incentivize regenerative production and nutritious end products, but instead we have encouraged industrialization and poor-quality food. We have tolerated inhumane conditions for farmworkers and an out-of-control mental health crisis among farmers. And we have turned a blind eye to the environmental disaster that is modern agriculture, a disaster enabled largely by government.

Thankfully, the policy situation began to shift under the Biden administration. The Inflation Reduction Act of 2022, for example, boosted spending on historically under-resourced federal and state programs, like the Conservation Stewardship Program and Environmental Quality Incentives Program, that fund regenerative and other forms of climate-smart agriculture.[8] Another government program for regenerative agriculture is the 2022 USDA Partnerships for Climate-Smart Commodities initiative, a more than $3.1 billion investment in projects that expand markets for climate-smart commodities, leverage the greenhouse gas benefits of climate-smart farming, and provide direct, meaningful benefits to production agriculture, including for small and underserved producers. For food processors, the USDA started a Resilient Food Systems Infrastructure Program to build out regional and local foodways and a $1 billion investment in expanding diversified, independent meat and poultry processing.[9]

And with young women like Vanessa involved, there is hope that the next Farm Bill will build on this progress.

Because the Farm Bill addresses so many issues, it brings together a diverse group of stakeholders and advocates. Climate activists might lobby for incentives that reduce farm emissions. Coalitions on racial justice might push for more resources for BIPOC farmers. The goal is to persuade members of Congress who sit on the Senate Committee on Agriculture, Nutrition, and Forestry and the House Committee on Agriculture to include their group's ideas in the Farm Bill. The House and Senate each pass their own version of the Farm Bill. The two bills go to a small conference committee of representatives and senators who merge the legislation into one compromise package, which is then debated on the two floors, amended as needed, and eventually passed by both chambers and signed by the president.[10] Or so the process is supposed to work.

On one hand, lawmaking is straightforward enough for *School-house Rock* to condense into its memorable "I'm Just a Bill" cartoon. On the other hand, the political environment is hyper-partisan as I write these words in 2023, a year in which a divided Congress must craft a new Farm Bill to replace the one expiring in October. Yet from a third perspective, the timing is also an opportunity. Regenerative agriculture and climate change mitigation are becoming bipartisan issues. More women serve in Congress and in positions of power within organizations that sway policy decisions. And the public is demanding government action on climate.

A 2023 "Climate Change in the American Mind: Beliefs and Attitudes" survey from Yale University and George Mason University found that about two in three Americans (65 percent) say the issue of global warming is either "extremely" (13 percent), "very" (22 percent), or "somewhat" (30 percent) important to them personally.[11] The same study also found that Americans who think global warming is happening outnumber those who think it is not happening by a ratio of more than 5 to 1 (74 percent happening versus 15 percent not).[12] A 2022 survey by the same researchers focused on politics and policy in particular found broad bipartisan support (honestly, such resounding support I was shocked) for pollution reduction, conservation, clean energy, and declaring climate change a national emergency.[13] More convincing evidence that climate is a bipartisan issue: in the red state of Iowa, 81 percent of farmers now believe climate change is happening, compared to 68 percent in 2011. More than half are concerned about climate impacts on their farms, up from 35 percent.[14]

In addition to voters, lawmakers consider the priorities of political groups that represent them, like Young Farmers. In the months before our interview, Vanessa met with congressional members' staff to learn their bosses' priorities for the bill and to persuade

them to consider Young Farmers' concerns. The deadline to submit language that could make its way into the bill is a few weeks
from my DC visit. The senators, representatives, and staff involved
with the Farm Bill are always busy, which adds an extra challenge
to her work. "We need to make sure that we make their life easier,
always," Vanessa says. "That means us writing the language to be
included in marker bills, us doing the research, us having a list
ready of who will like this bill, who won't like it. I always say, we
don't work *on* the Hill, we *work* the Hill."

The Young Farmers' offices are in the Methodist Building across
from the Supreme Court. While the location is impressive, and
convenient for meeting with congressional affiliates, the rooms
themselves are dated and cramped—perhaps a nod to cost consciousness. Vanessa leads me to a conference room where I will
play fly on the wall for a hybrid in-person/remote conversation
between representatives from the National Sustainable Agriculture
Coalition, the Rural Coalition, and Young Farmers. Since some
mission overlap exists between these groups, they are looking for
ways to work together rather than disparately or against one another on issues related to the Farm Bill. "We're all friends and need
to be talking to each other," Vanessa points out. One can bet Big
Ag and Big Food lobbyists are conducting meetings just like this
one to advance their agendas, so sticking together when possible
is important, especially for comparatively smaller, newer groups
like Young Farmers. "We are really young in the ecosystem of
farming organizations," Vanessa says. "We are only thirteen years
old. Compared to the Farm Bureau, which has been around, like,
a hundred years, or the National Farmers Union, we're still really
young."

Young Farmers is made up of local chapters around the nation.
Headed by farmer-members, these chapters work for racial justice,

build strong communities, and use policy and advocacy as tools for change. A team of staff and organizers across the country lead Young Farmers' work at the national and regional levels, overseen by a board of directors that is both majority BIPOC and majority farmer. Obviously the organization aims to represent the interests of its members, and young and diverse U.S. farmers in general— but how do leaders know what those interests are? "The way we do it is really participatory," Vanessa tells me. Young Farmers' heart is grassroots organizing. Every five years, Young Farmers surveys its members and affiliates to assess what ambitions, challenges, and policy solutions they would like to see enacted. The 2022 survey was the largest yet, with more than ten thousand responses.

"That survey really informs what we do. We gather the information, and then we will make policy recommendations based on that information," Vanessa says. With just six staff members working on federal policy change in DC, Young Farmers must be strategic with its lobbying; addressing all issue areas is not realistic. Instead, they focus on a smaller number of achievable outcomes. "There are many things we are going to fight for softly, but they are not going to be a priority," Vanessa says. Based on the 2022 survey, land access and conservation are top young farmer concerns, so those are the organization's main objectives for the Farm Bill.

Expanding land access and conservation incentives for young farmers are two areas where Young Farmers, the National Sustainable Agriculture Coalition, and the Rural Coalition might craft a unified pitch to lawmakers drafting the Farm Bill, Vanessa says. And this cycle is critical for attracting and keeping young farmers in the food system, because major changes are on the horizon in farm country. The average farmer age is about fifty-eight years. The last Census of Agriculture claims this figure represents "a long-term trend of aging in the U.S. producer population."[15]

More than half of U.S. farmland belongs to farmers fifty-five and older—and according to the Center for Rural Affairs, half of all current farmers are likely to retire in the coming decade.[16] Land is about to change hands, which could be an opportunity for young and would-be farmers.

Assuming young and would-be farmers materialize: since 1987, the number of entry-level farmers has fallen by 30 percent.[17] We currently have 321,261 total young farmers, defined as producers aged thirty-five or younger, and they comprise only 9 percent of the nation's 3.4 million farmers.[18] These numbers reflect the dismal realities of conventional agriculture as a business and lifestyle. Industrial farming is not attractive to the next generation. It is expensive, risky, and, in the eyes of a more ecologically conscious generation, too environmentally damaging. Newcomers hoping to farm on a smaller scale, or to farm sustainably, or even to join the conventional system often run into problems with land, which is hard to access and afford. It's no wonder that the number of young producers has contracted so much in the last few decades.

But with help, new young farmers could replace an aging generation, and existing young farmers could expand their operations. The moment is especially critical for BIPOC farmers, a group still underrepresented compared to their white counterparts but whose ranks are growing (see chapters 1 and 2 for the data). These groups currently struggle the most to obtain land.[19] Female and LGBTQ+ farmers, too, confront obstacles. If young farmers can't access land or succeed once they are on it, though, then industrial operators will likely acquire more acres, foreign buyers and nonagricultural entities will step in, and developers will encroach. Creating opportunities for prospective and current young farmers from all backgrounds offers the best chance of avoiding those outcomes.

What's more, an influx of new and young farmers could revitalize rural areas destroyed by agribusiness and farm consolidation.
Bringing those farmers to the land, people who have or likely will
have children, would ignite population growth and economic
renewal in small and medium-size towns—towns where today the
next generation leaves for the economic and social opportunities in
cities, as I did. Instead of shrinking, our rural areas could regenerate. Diversifying rural white communities would be an important added bonus, as many of the country's aspiring young farmers
are women, people of color, LGBTQ+, or immigrants. "The next
generation of farmer is a woman of color that is queer in an urban
area," Vanessa says. "That's what a lot of the consensus is and actually that's what a lot of the surveys miss." In Vanessa's observation I
see hope for rural communities that could benefit enormously from
diverse perspectives and cultural influences, and for new and young
farmers who could build their own economic security.

Young newcomers also are likely to undertake regenerative and
other forms of climate-smart agriculture if given the chance. A
look at the 2022 National Young Farmer Survey confirms this.
Eighty-six percent of young farmers who responded reported using
regenerative practices and 83 percent said conservation or regeneration is a primary purpose for their farm's existence—and that
is *without* robust institutional and social support for regenerative
agriculture. Given that 73 percent of respondents reported at least
one climate impact to their farm in the past year, they do not need
convincing that climate change is a threat.[20] "There is this idea of
land security and climate anxiety. They cannot have secure operations because they know they can be wiped out," Vanessa explains.

Young farmers understand that one way to mitigate climate
threats is via sustainable agriculture. Getting more of these young
producers on the land could help make regenerative agriculture

mainstream and protect rural areas from climate impacts. Regenerative farms are also more productive because of their diversification and thus can be smaller acreage-wise, making room for more families and farmworkers on the land. Strengthening rural communities and decentralizing agricultural production both work to stabilize the food system. "For us to put farmers on the land who care about conservation practices is both protecting the American food supply chain and fighting climate change," Vanessa notes.

Is it hard to convince legislators and their staff to hear and act on those messages? I ask Vanessa. "Not the friendly ones," she says with a laugh. "I think it's about challenging narratives. Some people do not understand our farmers and a lot of what this generation wants to do. These farmers, they do not care only about making a profit. They care about protecting the soil and feeding their communities. Yes, they will do a business plan and apply for credit, but they do not care just about making money. They care about having a dignified livelihood that allows them to do their part to fight climate change and fight hunger." Once again, stereotypical and outdated perceptions about farmers hinder the progress of regenerative agriculture. Lawmakers, many of advanced age or personally removed from agriculture, may not recognize that young farmers are not carbon copies of the previous generation that champions industrial agriculture. Neither is the new cohort as white or male.

On the other side, young farmers may not recognize that today's USDA is not yesterday's USDA. Many young producers, especially BIPOC and female farmers, distrust the USDA (see chapter 2 for a refresher on why). Vanessa says the majority of Young Farmers' members do not utilize USDA programs and express skepticism about the agency when polled. But like it or not, the USDA carries out most Farm Bill directives and serves as the main access point for farmer assistance. Vanessa believes that repairing the relationship between young farmers and the federal government is integral to those producers' success. Otherwise, they could miss out on critical support.

And while she encourages private entities to help young farmers, giving up on government and outsourcing young farmer support to businesses, nonprofits, or other organizations is not productive. Doing so would hand the government—by far the largest source of aid to farmers—a free pass to ignore its responsibilities, and also confirm to young farmers that the government does not care about them unless they fit the white, male, industrial agriculture mold. "We are trying to rebuild that trust that these institutions should work for you," she says. "I think for me that's why I always call what I do civic agriculture. It's the idea that agriculture is the way we engage with government, at least as farmers and as eaters. These institutions should be working for us and for the farm and food future we're building, not just for the corporations and commodity farmers."

Rural sociologist Thomas Lyson coined the term "civic agriculture" and defined it as "the embedding of local agricultural and food production in the community. Civic agriculture is not only a source of family income for the farmer and food processor; civic agricultural enterprises contribute to the health and vitality of communities in a variety of social, economic, political, and cultural ways. . . . [It] embodies a commitment to developing and strengthening an economically, environmentally, and socially sustainable system of agriculture and food production that relies on local resources and serves local markets and consumers."[21] Civic agriculture sounds much like traditional Indigenous, African, Black, Latinx, and Asian farming systems; it is a way of life and food production that supports individuals, communities, and the environment simultaneously. When Vanessa describes her advocacy as civic agriculture, she recognizes how young farmers fit within, and contribute to, local agricultural systems—systems that can give rise to the social, economic, political, and cultural benefits Lyson envisions if we base them on regenerative practices.

Voters want an agricultural system that looks more like the civic agriculture Lyson describes. A National Farm Bill Poll conducted in March 2023 by the John Hopkins Center for a Livable Future found that 90 percent of respondents favor regenerative agriculture, and an even higher percentage believes the federal government should help small and midsize agricultural producers. The poll revealed virtually equivalent support for sustainable food production that ensures healthy food now and in the future. Seventy percent of respondents affirmed that they would pay more for sustainably produced food, and 75 percent would do the same for meat from animals raised on smaller, family-owned farms instead of confinements.[22]

Spending data affirms these responses. Organic food sales totaled $56.4 billion in 2020, an increase of 12.8 percent (compared to 4.6 percent growth in 2019), according to the Organic Trade Association's 2021 Organic Industry Survey. Past OTA surveys found that millennials are driving steady increases in organic sales over time. In 2020, the pandemic fueled a record-breaking jump in sales that cooled slightly in the following years, but nevertheless demonstrated undeniable continued support for organic. Today's shoppers do not want antibiotic-laden meat, pesticide-covered produce, or processed foods with genetically modified ingredients, and those attitudes held true even during a time of crisis. The modern consumer is looking for quality measured in terms of health, environmental impact, and benefits to farmers and farmworkers—priorities that voters deserve to have reflected in the Farm Bill.

The Farm Bill is important—critical even—but it and other federal programs, laws, and regulations are not the only drivers of food and agriculture change. States shape policy as well. For example, in early 2023 California's Department of Food and Agriculture and State Board of Food and Agriculture announced their Ag Vision for the Next Decade plan. The plan guides the state's efforts in five key areas: fostering climate-smart, resilient, and regenerative food systems; building healthy, local communities; driving next-generation talent and tools; enhancing understanding of agriculture; and collaborating on smarter regulations.[23] That same year Washington state lawmakers passed a law targeting food waste in a number of ways, including the creation of the Washington Center for Sustainable Food Management.[24] In 2022 the Minnesota Department of Agriculture directed USDA funds to schools so they could purchase more local food and establish partnerships with underserved farmers and small businesses.[25] States can influence food decisions in other institutions, like colleges, health care facilities, prisons, and government offices.

Individual cities direct food policy as well. The city of Baltimore provides tax credits to supermarkets that locate or renovate in food deserts and meet requirements for healthy food offerings, and also encourages urban agriculture by relaxing rules around animal husbandry and hoop houses.[26] Cities and counties across the United States have instituted soda and other junk-food taxes to improve health, mandated health and sourcing requirements for schools and other city entities, and boosted support for farmers' markets.

These efforts demonstrate the need not only for holistic food and agriculture policies at all levels, but also for leaders like Vanessa to cultivate those policies, often without a thank-you. The reality is that such leadership often unfolds behind the scenes. The public tends to be interested in farmers implementing regenerative agriculture—think films like *The Biggest Little Farm* and *Kiss the Ground*—but elected officials, nonprofit workers, academics, and so forth are equally essential in building a regenerative food system, even if their work isn't terribly visible. Today's hyper-partisan political climate and the slow-moving nature of government also dampen enthusiasm for institutional change. "It's really hard for some farmers to understand that this game is worth playing," Vanessa says. "A lot of what we're doing is moving needles little by little, changing language year by year. A lot of people don't see that is going to transform the food system. But if you compare the food system that we had twenty years ago with the one we have today, we have way more investment in local food, regional and sustainable agriculture, because we have been moving the needle in every Farm Bill and every appropriation cycle."

After the joint stakeholder discussion, Vanessa and I head out for a major meeting with staffers representing Democratic members of the House Committee on Agriculture (Young Farmers is nonpartisan and meets with both parties, by the way). She and I walk between the Supreme Court and Capitol toward the Longworth House Office Building. The afternoon is 47 degrees and sunny with a brisk eighteen-mile-per-hour wind that suggests winter even though the pink saucer magnolias are in full bloom. I won't be allowed in the meeting, so we stroll and talk.

Vanessa tells me about her recent appointment to the Rhode Island USDA Farm Service Agency (FSA) state committee. "It is the biggest, scariest, most competitive thing I've ever applied for," she says. The FSA administers credit and loan programs, as well as conservation, commodity, disaster, and farm marketing programs. FSA state committee members oversee the agency's programs and county committee operations within their state, resolve program delivery appeals, maintain relations with industry stakeholders, and inform producers about FSA programs. Serving the FSA, a federal agriculture entity, while also advocating on behalf of agricultural producers via Young Farmers puts Vanessa in a strange but useful position. "It makes me so much better at my job. I'm getting the inside and the outside at the same time," she says.

It's not the first time she has experienced being inside and outside of a system at the same time. Vanessa grew up in the Dominican Republic's Cibao Valley, a region she calls "the Midwest of the Dominican Republic" that is dominated by crops like cassava, plantains, and sweet potatoes grown in monoculture. She and her family moved to Rhode Island in 2011 when she was fifteen years old. At the time, her new home of North Providence seemed far less integrated into a local or regional food system than her agriculture-focused hometown of Moca was back in the Dominican Republic. "The community food system was just emerging when I arrived to the United States. Farmers' markets were popping up and things like that. I was really interested in green businesses and how businesses can fight climate change and contribute to sustainable development," she recalls.

Driven by those ideals, Vanessa emerged early on as a sustainable food leader. North Providence's mayor named her manager of the town's farmers' market while she was still in high school. Inspired by her agricultural exposure both in the States and in the Dominican Republic, Vanessa sought an undergraduate degree in environmental and natural resource economics from the University of Rhode Island and later a master's degree in community food and agricultural systems from Michigan State University. She completed multiple university and government research assistantships and worked at the USDA National Organic Program and the Michigan State University Center for Regional Food Systems. A few other highlights from her résumé: 2019 Cynthia Hayes Scholar, 2019 James Beard Foundation National Scholar, 2020 Michigan Junior

Food System Leader of the Year, and 2021 Emerging Food and Agriculture Leader. She holds a seat on the National Sustainable Agriculture Coalition's Organizational Council and has served on its Farming Opportunities and Fair Competition committee in the past.

These roles eventually led Vanessa to Young Farmers in April 2020. She started as federal policy associate and was promoted through the ranks to her current position as director of government relations. Vanessa is passionate about young farmers and their success, of course, but her service to them is also an intentional career choice. It's hard to make change or build coalitions when your focus is too broad, she explains, so zeroing in on one group or a narrower set of issues can be more effective. "Something that I have always been really afraid of is being a mile wide and an inch deep. I would rather be good at something than be in many spaces at the same time and not make a change," she says. "That's why I went really focused on ag, and went more focused on sustainable ag, and now even more focused on young farmers, when before it was more local food and sustainable ag."

Sustainability is a common thread running through Vanessa's career, and her definition of the term has evolved over the years. "I think my model is definitely much more human-centric," Vanessa told me via Zoom when I asked earlier what sustainability means to her. "When I look at how previous conversations about the food system started, it was really more about what we ate and where our food came from. It was never about a lot of the power dynamics or who gets to participate in that system. I think now we're getting closer to understanding that sustainability is not just about consumer choices, or about food, or environmental outcomes. It's about people."

In other words, sustainability means ensuring that everyone can participate in and benefit from the food system. This, I think, is another example of what Kathryn Brasier, professor of rural sociology at Pennsylvania State University, meant when she told me that infusing more creativity into the food system through diversity and inclusion makes us more likely to solve the system's many problems. Vanessa's Dominican heritage, gender, and immigrant experience are just a few reasons her leadership is effective and valuable, especially in a warming world that requires diverse perspectives.

Vanessa may epitomize the changing face of agricultural leadership, but her presence also reinforces lingering representation gaps—creativity lost, so to speak, and with it reduced chances of successfully navigating a hotter planet. While she says women of color are better represented in community organizing, they are the exception, not the norm, in the agriculture policy world. "I know as a young woman of color that is an immigrant, I have the privilege of having the role that I do, especially interacting with lawmakers, helping design policy, and implementing program management and budgetary decisions for USDA, and other roles that I have," Vanessa says. "Seeing that, I'm always like, 'Why are there not more people like me here?' Immigrants work all across the food system, and so many of us don't get the opportunity to actually impact the governance of the food system and shape the food system. Obviously there's a thousand barriers, from economics to language to power dynamics to actually knowing how to get a seat at the table. So I always feel really privileged, but I see this privilege as a right and a responsibility."

The typical person of authority in the agriculture world is white, male, and older, with a background in large-scale conventional farming, Vanessa says. If Vanessa isn't breaking this mold for agriculture leadership, then I don't know who is. She has never farmed. She's a female immigrant of color. She's young, just twenty-seven when I meet with her. All this is not lost on her. "When I seek opportunities that I know I'm qualified for and my community supports me, but it has never been given to a young woman of color, it's like, 'This is your chance to change the recipe, the formula.' That takes a lot of courage from the institutions and from ourselves to put ourselves forward because it's a lot of harm that you're putting yourself through as a person of color, a lot of vulnerability that you have to be ready to hear and walk through and handle," she says. Courage is definitely a requirement for leadership, doubly so when a would-be leader's gender, race, sexual orientation, or age stands out. It's an unfair burden to carry, and one we ought to eliminate with inclusivity.

I end my visit with Vanessa in the most fitting of ways: a meal. Food has brought us together, so it makes sense that food becomes our way of parting for now. We head to a bustling French café near the Capitol. Sitting down with our meals, we are two ordinary eaters partaking in a food system we are both trying to heal in our own ways, with our unique gifts and backgrounds. We are women, and we are young, and we are not giving up. "I know that my role in the ecosystem of food justice and young farmers is making government work for young farmers and giving them hope that these institutions can work for them, and also holding those institutions accountable," Vanessa says. "It took me a long time to get here and hopefully I stay here because I still have the urge to do a thousand things."

Conclusion

I finish this book where I started: the mountains of western North Carolina. This time it is the heart of summer. The pandemic is over, but there are new emergencies. A long-lasting, deadly heat wave bakes the southern United States. Temperatures in Phoenix, Arizona, reach 118 degrees and the city spends thirty-one consecutive days at 110 degrees or higher. "Feels-like" temperatures in South Florida climb above 115. Meanwhile, uncontrollable forest fires rage across Canada. Smoke smothers the Midwest and Northeast and large swaths of the South. When my husband and I take visitors to Mount Mitchell, the highest peak east of the Mississippi, a haze blankets the mountains. The smoke is almost like fog; we can barely see the nearest ridges. A few weeks later floods strike the Northeast and ocean temperatures in the Florida Keys hit 101 degrees, the hottest ever recorded. Corals bleach and die en masse.

Climate change and how it fuels these extremes is in the headlines daily. It's hard not to feel a sense of dread, like at any moment disaster could strike, because it increasingly might. I find myself recalling the late Barry Lopez's question, or read another way his call to action, when considering the changing planet: "In this moment, is it still possible to face the gathering darkness and say to the physical Earth, and to all its creatures, including ourselves, fiercely and without embarrassment, I love you, and to embrace fearlessly the burning world?"[1]

Despite all the climate chaos (and also because my book deadline looms), food is on my mind more than disaster. And in a pleasant coincidence, food becomes a central focus of my non-writing hours, too, as North Carolina offers up its abundance in the form of area-grown products delivered through localized foodways. Here's an example: last summer, a kind neighbor stopped by and gifted my husband and I with several bags of produce and a massive watermelon. He volunteered with Dig In! Yancey Community Garden, he explained, a nonprofit with a two-acre farm that, together with its network of sister gardens and other local farms, offers free (yes, free) shares of produce and eggs every week at its walk-up market. Dig In! uses organic regenerative practices, trains young and beginner growers, channels its output to other area food programs, and assists fellow community gardens in getting started and thriving. The nonprofit "envision[s] a vibrant local food system, built from collaboration and community care, where everyone has access to plentiful, locally grown fresh food. By coming together and sharing the work, resources, knowledge, and fruits of the harvests, the fabric of our community is strengthened."

This summer, the same neighbor comes by with a bag or two of produce a week. He also brings blueberries, the best I've ever tasted, from the patch outside his house. Every day my husband and I eat something or other from the bounty: collards, beans, cucumbers, kale, fennel, yellow squash, Swiss chard, lettuce, spinach, mint, basil, zucchini, radishes, onions, potatoes, eggs. We also frequent a farm on the outskirts of town that sells its own produce and locally raised extras such as beef, pork, and eggs at its store. The place aggregates regional luxuries like Georgia peaches and melons, South Carolina strawberries, and Tennessee tomatoes as well. Then there's the weekly town farmers' market with cheese, bread, baked goods, grits, honey, jams, more produce, and all manner of meats. What's more, restaurants from here to Asheville and beyond showcase local and regional products. I've never eaten so well in my life.

This six-week period is the first time I have relied primarily on local food systems for sustenance. Yes, I still shopped at grocery stores occasionally. But I caught a glimpse of what is missing in the food system, not just in my home base of South Florida but across the nation. I also saw what was possible. In this corner of western North Carolina, a community built an equitable, nutritious, locally focused food system in spite of the forces working against it. It continually has to fight off Walmart, Dollar General, and fast food. Young people tend to leave for cities, and the economy is fragile. Opioid addiction and poverty are painfully visible. The difference seems to be that people here care enough to create the food system they want, and a diverse group of nonprofits, churches, government actors, volunteers, and farms provide scaffolding for that effort. In Dig In!'s case, they have financial support from grants and other sources and meaningful community partnerships. They are committed to regenerative organic production. People are working together, not alone. Even when the farmers' market closes for the winter, local stores carry canned items, frozen meat, baked goods, and other area specialties. Not everyone in the community seeks out local and regional food—but it is a realistic, nutritious, and reliable option for those who want it. They've created a system rooted in empathy.

Perhaps what that community has done is one way to embrace fearlessly a burning world, as Kelsey, Carrie and Erin, Susan, and Josh and Jordan do on the land every day. As Bu, Karen, and Mary Jane do by leading a company for people and planet, not profit. As Tina and Vanessa do in advocating and organizing. As Wen-Jay does in connecting farmers and consumers, and as Sarah and Esther do by driving regenerative investment. Perhaps we all have ways of embracing our on-fire world through such solutions-oriented work.

But as much as these women leaders and the regenerative movement they represent inspire hope, I still ask myself whether humanity can summon the collective will to enact regenerative thinking writ large. We've long held blueprints for social and environmental change—land conservation, clean energy, universal health care, robust public education—and failed to act. It's hard to imagine us acting differently in response to climate change. But as Robin Wall Kimmerer writes, despair is a useless feeling to which we cannot submit. Our only option is to restore what we have destroyed: "Restoration is a powerful antidote to despair. Restoration offers concrete means by which humans can once again enter into positive, creative relationship with the more-than-human world, meeting responsibilities that are simultaneously material and spiritual. It's not enough to grieve. It's not enough to just stop doing bad things."[2]

It's not enough to just stop doing bad things. We must do good things instead. She is speaking about the natural world, but I think her words apply more broadly. Restoration is not cessation, but action. That is the restorative promise of regenerative agriculture and a regenerative food system, rooted in the BIPOC wisdom of which Kimmerer is a part.

No single person or action featured in this book—regenerative ranching or farming, researching, advising, capitalizing, policy-making, advocating, creating local and regional supply chains—will fix the food system. But I believe collective action in those areas and beyond can. Collective action does not mean everybody does the same thing. It means we serve a mutual goal, each in our own way, through the lens of shared ideals and cooperation. We communicate. As A-dae Romero-Briones, director of programs—Native Agriculture and Food Systems at First Nations Development Institute, explained to me, we need conversations between areas of specialization as we return to a regenerative food system based in Indigenous practices and concepts. Isolating components of the food system, or prioritizing certain parts or people as more important, means we lose sight of how regenerative thinking fosters health within the whole ecosystem. She uses "ecosystem" as a broad term for relationships between humans and land, between history and future, between people. "Reducing the conversation about regenerative agriculture to just carbon or just soil health hides all the complex and interconnected benefits of a regenerative system," Romero-Briones says.

Vanessa introduced me to another ecosystem-based way to understand what collective action looks like. Over lunch in DC that day, she referenced the ideas of Deepa Iyer, whose latest book is *Social Change Now: A Guide for Reflection and Connection.* The book builds on the social change ecosystem map Deepa created through her work at the Building Movement Project, a national nonprofit organization that catalyzes social change through research, relationships, and resources. In Deepa's ecosystem framework, she explains ten common roles people tend to play in advancing social change: weaver, experimenter, front-line responder, visionary, builder, caregiver, disrupter, healer, storyteller, and guide. No role is more or less important than another. There is no single or "right" way to promote change, and people may embody more than one role.

Thinking about social change as an ecosystem in which a diverse group of changemakers work together while also embracing their individual roles does two things. First, it helps people appreciate how their individual contributions, however small, contribute to larger goals—which can help prevent a feeling of futility or burnout. For example, experimenters see how their push to develop new methods or practices is important for builders who implement them. Second, it reinforces the importance of connections and shared values. In ecosystem thinking, social change is not about hierarchies or one-dimensional approaches. It is about collaboration and holistic thinking.

The ecosystem approach is well suited—I would even say required—for the enormous task of building a regenerative food system. We need people to play all ten of Deepa's social change roles on every link of the food chain. Within those links we want partnership instead of competition. Common values should guide our construction of a regenerative food system—and this time health, not profit, is a central value. Healthy soil, crops, livestock, and environments. Healthy consumers, families, workers, and communities. Healthy relationships that promote diversity, equity, and inclusivity. Healthy agricultural and business practices that lead to sustainable, dignified livelihoods. Policies and programs that encourage health in all its forms.

The stories in these pages are examples of the efforts and ideas we need, not from women alone but from people of all genders and backgrounds, all ages and abilities. Change is already well underway but could fall short if we do not rapidly expand our efforts. We need more changemakers and thought leaders, and more diversity in both groups. We need more government and private sector support, more voters and consumers involved. Above all, we need more urgency. Women provided much of the energy that put the regenerative agriculture movement in motion. It's up to all of us to tap into our individual energy, identify our role in the movement, and work toward a regenerative, resilient food system to help save ourselves, our future generations, and our planet.

Acknowledgments

Collaboration is more than a key theme of the regenerative agriculture transition—it is why this book exists. So many people gave of their time, knowledge, and energy to help me understand (or try to!) the complex ideas and processes involved in building a sustainable food system. What the women featured here told me about the regenerative agriculture and food world is true in my experience: people are eager to assist and lift up one another, and I appreciate every person who said yes when I asked for help. Any mistakes or omissions in this book are entirely mine.

Many thanks to the experts who gave in-person and remote interviews: A-dae Romero-Briones, Amanda Zakharov, Cathy Geary, Claudia Carter, Dawn Hoover, DeAnn Presley, Gabrielle Roesch-McNally, Gail Fuller, Lynnette Miller, Jessica Hulse Dillon, Jill Clapperton, Kathryn Brasier, Kelly Clark, Lisa Kivirist, Maia Hardy, Maritza Pierre, Mary Hendrickson, Russ Schumacher, Sarah Harper, Sariah Hoover, and Virginia Harris. Special thanks to Cherilyn Yazzie, Holly Whitesides, Dreu VanHoose, Mary Cory, Tannis Axten, and Liz Haney—your stories may not be here, but they informed my thinking tremendously. Thank you for sharing. Same to Linda Pechin-Long and Keith Long; I appreciate your friendship and generous hospitality so very much. Thank you, Jessica Martell, Sarah Beth Hopton, Blue Ridge Women in Agriculture, and Carol and Lon Coulter for hosting me in North

Carolina. To John Blair at Konza Prairie Biological Station and Tim Crews of the Land Institute, thank you for educating me and showing me around. I also appreciate the scientists, investigative journalists, and thought leaders I've never met but whose work I consulted in writing this book.

I am tremendously, eternally thankful to the following people who shared their stories in such honest detail, who spent precious time talking with me and hosting me, who offered valuable feedback on my writing, who bravely allowed readers to see their role in creating a regenerative food system. Kelsey Scott, Carrie Martin, Erin Martin, Susan Jaster, Josh Payne, Jordan Welch, Bu Nygrens, Mary Jane Evans, Karen Salinger, Wen-Jay Ying, Tina Owens, Sarah Day Levesque, Esther Park, and Vanessa García Polanco—thank you for *everything*. Your leadership and talent inspire me to keep doing whatever I can to make regenerative a reality. I owe this book to each of you.

Thank you to the institutions that generously funded this project over the years: Ninth Letter and the Illinois Regenerative Agriculture Initiative, the Money for Women/Barbara Deming Memorial Fund, Florida Atlantic University's Dorothy F. Schmidt College of Arts and Letters, and the Richard J. Margolis Award. Many thanks to my colleagues and students in Florida Atlantic's English Department for their unwavering support and encouragement. I am grateful for the expertise of Amanda Annis at Trident Media Group; Marc Favreau, Rachel Vega-Decesario, and zakia henderson-brown at The New Press; copy editor extraordinaire Eileen Chetti; and everyone involved with the publishing, distribution, and marketing of this book. I leaned on many other people during this project and related career endeavors, such as Monica Schneider, John Jeep, Ayşe Papatya Bucak, Andrew Furman, Becka McKay, Brittany Ackerman, and Patrick Hicks. Thank you to my amazing friends near and far who asked after the book, listened when I was frustrated, and cheered me on. My extended family, particularly Dary and Deb Peckham and Jamie and Matt Fisher, always did the same—thank you so much.

I can't quite convey how deeply thankful I am for my siblings Anna Johnson and Jeremie Papon, Joshua Johnson, and Charlotte Johnson, and my parents, Les and Cathy Johnson. How do you properly thank people who love you continually, believe in you (despite frequent evidence to the contrary!), and inspire you with their own achievements? Having such incredible family is a gift I don't deserve but am so, so grateful to have received. I feel the same way about Ryan Anderson, my husband, best friend, and steadfast champion. He helped me navigate the book's highs and lows, as he does so deftly in our life generally. I would be lost without this most wonderful partner a person could have.

Notes

Introduction

1. Hanh Nguyen, "Sustainable Food Systems: Concept and Framework," Food and Agriculture Organization of the United Nations, 2018, 1.

2. Claire Kelloway and Sarah Miller, "Food and Power: Addressing Monopolization in America's Food System," Open Markets Institute, September 21, 2021, 8.

3. Kelloway and Miller, "Food and Power," 11.

4. "The Economic Cost of Food Monopolies: The Grocery Cartels," Food and Water Watch, November 2021, 1.

5. Kelloway and Miller, "Food and Power," 10.

6. Jessica Fu, "Can $1 Billion Really Fix a Meat Industry Dominated by Just Four Companies?," *The Counter*, January 5, 2022.

7. Kelloway and Miller, "Food and Power," 3–4.

8. "Economic Cost of Food Monopolies," 5.

9. James M. MacDonald, Robert A. Hoppe, and Doris Newton, "Three Decades of Consolidation in U.S. Agriculture," USDA Economic Research Service, *Economic Information Bulletin* 189, March 2018, 21–25.

10. U.S. Department of Agriculture National Agriculture Statistics Service, "2017 Census of Agriculture Highlights: Farm Economics," April 2019.

11. "Economic Cost of Food Monopolies," 2; Kelloway and Miller, "Food and Power," 9.

12. "Facts About Agricultural Workers," National Center for Farmworker Health, January 2022, http://www.ncfh.org/facts-about-agricultural-workers -fact-sheet.html.

13. "Slavery in the U.S.," Food Empowerment Project, https://foodispower .org/human-labor-slavery/slavery-in-the-us/.

14. Michael Grabell, "The Plot to Keep Meatpacking Plants Open During COVID-19," *ProPublica*, May 13, 2022.

15. U.S. Bureau of Labor Statistics, "Consumer Price Index: 2022 in Review," *TED: The Economics Daily*, January 17, 2023, https://www.bls.gov /opub/ted/2023/consumer-price-index-2022-in-review.htm.

16. Isabella Simonetti and Julie Creswell, "Food Prices Soar, and So Do Companies' Profits," *New York Times*, November 8, 2022.

17. Philip H. Howard and Mary Hendrickson, "Corporate Concentration in the US Food System Makes Food More Expensive and Less Accessible for Many Americans," *The Conversation*, February 8, 2021.

18. "Economic Cost of Food Monopolies," 2.

19. For a fuller understanding of how exactly climate change will impact agriculture and rural communities, see Carolyn Olson et al., "Agriculture and Rural Communities," in U.S. Global Change Research Program, *Impacts, Risks, and Adaptation in the United States: Fourth National Climate Assessment*, vol. 2, ed D. R. Reidmiller et al., chap. 10 (Washington, DC: U.S. Global Change Research Program, 2018).

20. Georgina Gustin, "As Extreme Weather Batters America's Farm Country, Costing Billions, Banks Ignore the Financial Risks of Climate Change," *Inside Climate News*, May 2, 2021.

21. Amanda Little, *The Fate of Food: What We'll Eat in a Bigger, Hotter, Smarter World* (New York: Harmony Books, 2019), 42–50.

22. Bradley L. Hardy and Trevon D. Logan, "Racial Economic Inequality amid the COVID-19 Crisis," Hamilton Project—Brookings Institution, August 2020.

23. Climate Reality Project, "How Feedback Loops Are Making the Climate Crisis Worse," January 2, 2020, https://www.climaterealityproject.org /blog/how-feedback-loops-are-making-climate-crisis-worse.

24. Dominik L. Schumacher et al., "Drought Self-Propagation in Drylands Due to Land–Atmosphere Feedbacks," *Nature Geoscience* 15, no. 4 (April 2022): 262–68.

25. A. Park Williams, Benjamin I. Cook, and Jason E. Smerdon, "Rapid Intensification of the Emerging Southwestern North American Megadrought in 2020–2021," *Nature Climate Change* 12 (March 2022): 232–34.

26. David I. Armstrong McKay et al, "Exceeding 1.5°C Global Warming Could Trigger Multiple Climate Tipping Points," *Science* 377, no. 6611 (September 9, 2022).

27. S. G. Potts et al., "Summary for Policymakers of the Assessment Report of the Intergovernmental Science-Policy Platform on Biodiversity and Ecosystem Services on Pollinators, Pollination and Food Production," Intergovernmental Science-Policy Platform on Biodiversity and Ecosystem Services, 2016, 8.

28. S. Diaz, "Summary for Policymakers of the Global Assessment Report on Biodiversity and Ecosystem Services of the Intergovernmental Science-Policy Platform on Biodiversity and Ecosystem Services," Intergovernmental Science-Policy Platform on Biodiversity and Ecosystem Services, 2019, 12.

29. Christopher J. Rhodes, "Are Insect Species Imperiled? Critical Factors and Prevailing Evidence for a Potential Global Loss of the Entomofauna: A Current Commentary," *Science Progress* 102, no. 2 (June 2019): 183.

30. Diaz, "Summary for Policymakers of the Global Assessment Report," 12.

31. Claire E. LaCannel and Jonathan G. Lundgren, "Regenerative Agriculture: Merging Farming and Natural Resource Conservation Profitably," *PeerJ* (2018): 1.

1. Disturbance

1. Tom Whipple, "'The Worst Thing I Can Remember': How Drought Is Crushing Ranchers," *New York Times*, August 29, 2021.

2. World Rangeland Learning Experience (WRANGLE), "North American Short Grass Prairie," https://wrangle.org/ecotype/north-american-short-grass-prairie.

3. John Blair, director of the Konza Prairie Biological Station, interview with the author, June 11, 2021.

4. Jonathan Proctor, Bill Haskins, and Steve C. Forrest, "Focal Areas for Conservation of Prairie Dogs and the Grassland Ecosystem," in *Conservation of the Black-Tailed Prairie Dog*, ed. John L. Hoogland (Washington, DC: Island Press, 2005), 236.

5. The idea that prairie dogs are bad for livestock production—a claim repeated so often that it has hardened into fact—melts into myth after a closer look. Here I drew from James K. Detling, "Do Prairie Dogs Compete with Livestock?" in *Conservation of the Black-Tailed Prairie Dog*, ed. John L. Hoogland (Washington, DC: Island Press, 2005), 74; John L. Hoogland, "Natural History of Prairie Dogs," in *Conservation of the Black-Tailed Prairie Dog*, ed. John L. Hoogland (Washington, DC: Island Press, 2005), 6; Rodrigo Sierra-Corona

et al., "Black-Tailed Prairie Dogs, Cattle, and the Conservation of North America's Arid Grasslands," *PLOS One* 10, no. 3 (2015): 6–7.

6. Blair interview.

7. Christopher I. Roosa et al., "Indigenous Impacts on North American Great Plains Fire Regimes of the Past Millennium," *Proceedings of the National Academy of Sciences* 115, no. 32 (July 23, 2018), https://www.pnas.org/doi/full /10.1073/pnas.1805259115.

8. Liz Carlisle, *Healing Grounds: Climate, Justice, and the Deep Roots of Regenerative Farming* (Washington, DC: Island Press, 2022), 20–21.

9. Jehangir H. Bhadha et al., "Raising Soil Organic Matter Content to Improve Water Holding Capacity," Document SL447, Department of Soil and Water Sciences, UF/IFAS Extension, September 1, 2021, https://edis.ifas.ufl .edu/publication/SS661.

10. National Park Service, "Last Stand of the Tallgrass Prairie," https://www .nps.gov/tapr/index.htm.

11. World Rangeland Learning Experience (WRANGLE), "North American Mixed Grass Prairie," https://wrangle.org/ecotype/north-american -mixed-grass-prairie.

12. WRANGLE, "North American Short Grass Prairie."

13. Gary Clayton Anderson, *Ethnic Cleansing and the Indian: The Crime That Should Haunt America* (Norman: University of Oklahoma Press, 2014).

14. Carolyn E. Sachs et al., *The Rise of Women Farmers and Sustainable Agriculture* (Iowa City: University of Iowa Press, 2016).

15. National Drought Mitigation Center, University of Nebraska, "How Can Overgrazing Leave Grass Vulnerable to Drought?," https://drought.unl.edu /ranchplan/DroughtBasics/GrazingandDrought/OvergrazingandDrought .aspx.

16. Claire Kelloway and Sarah Miller, "Food and Power: Addressing Monopolization in America's Food System," Open Markets Institute, September 21, 2021, 3–4.

17. Kelloway and Miller, "Food and Power," 3.

18. "Ranch Group Urges FTC, DOJ to Investigate Vertical Integration of Cattle Feedlots," *Tri-State Livestock News*, April 29, 2022, https://www.tsln .com/news/ranch-group-urges-ftc-doj-to-investigate-vertical-integration-of -cattle-feedlots/.

19. Carrie Hribar, "Understanding Concentrated Animal Feeding Operations and Their Impact on Communities," National Associa-

tion of Local Boards of Health, 2010, https://www.cdc.gov/nceh/ehs/docs /understanding_cafos_nalboh.pdf.

20. Stephan van Vliet, Frederick D. Provenza, and Scott L. Kronberg, "Health-Promoting Phytonutrients Are Higher in Grass-Fed Meat and Milk," *Frontiers in Sustainable Food Systems* 4 (February 1, 2021): 4–5.

21. This is not to say that I, as a white woman, can fully understand the Indigenous experience or the full history of settler colonialism, or to suggest that white people are not *still* complicit in the oppression of Native people. I mean to convey humility in being allowed onto the reservation and given the chance to learn from people like Kelsey.

22. Gosia Wozniacka, "Does Regenerative Agriculture Have a Race Problem?," *Civil Eats*, January 5, 2021, https://civileats.com/2021/01/05/does -regenerative-agriculture-have-a-race-problem/.

23. Tracy Heim, "The Indigenous Origins of Regenerative Agriculture," National Farmers Union, October 12, 2020, https://nfu.org/2020/10/12/the -indigenous-origins-of-regenerative-agriculture/.

24. Wozniacka, "Does Regenerative Agriculture Have a Race Problem?"

25. Mark Bittman, *Animal, Vegetable, Junk: A History of Food, from Sustainable to Suicidal* (Boston: Mariner Books, 2022), 31.

26. Carlisle, *Healing Grounds*, 98.

27. Carlisle, *Healing Grounds*, 140–42.

28. For more thoughts on the Indigenous philosophy of reciprocity, read Robin Wall Kimmerer's *Braiding Sweetgrass: Indigenous Wisdom, Scientific Knowledge, and the Teachings of Plants* (Minneapolis, MN: Milkweed Editions, 2013).

29. Kirk Siegler, "A mega-drought is hammering the U.S. In North Dakota, it's worse than the Dust Bowl," *NPR*, October 6, 2021, https://www.npr.org /2021/10/06/1043371973/a-mega-drought-is-hammering-the-us-in-north -dakota-its-worse-than-the-dust-bowl.

30. U.S. Drought Monitor, https://droughtmonitor.unl.edu/.

31. Katie Wedell, Lucille Sherman, and Sky Chadde, "Seeds of Despair," *USA Today*, March 9, 2020, https://www.usatoday.com/in-depth/news /investigations/2020/03/09/climate-tariffs-debt-and-isolation-drive-some -farmers-suicide/4955865002/.

32. Ronnie Cummins and Andre Leu, "Regenerative Grazing—Increased Production, Biodiversity Resilience, Profits and a Climate Change Solution," Regeneration International, March 29, 2021, https://regenerationinternational

.org/2021/03/29/regenerative-grazing-increased-production-biodiversity
-resilience-profits-and-a-climate-change-solution/.

33. Here I echo Robin Wall Kimmerer's comparison of the relationship between land and people to the three rows of a basket in traditional Indigenous basket weaving: "Ecological well-being and the laws of nature are always that first row. . . . The second reveals material welfare, the subsistence of human needs. Economy built upon ecology . . . By using materials as if they were a gift, and returning that gift through worthy use, we find balance. I think that third row goes by many names: Respect. Reciprocity. All Our Relations. I think of it as the spirit row. Whatever the name, the three rows represent recognition that our lives depend on one another, human needs being only one row in the basket that must hold us all." Kimmerer, *Braiding Sweetgrass*, 152–53.

34. USDA 2017 Census of Agriculture, "New and Beginning Producers," https://www.nass.usda.gov/Publications/Highlights/2020/census -beginning%20-farmers.pdf.

35. USDA 2017 Census of Agriculture, "Farm Producers," https://www .nass.usda.gov/Publications/Highlights/2019/2017Census_Farm_Producers .pdf.

36. USDA 2017 Census of Agriculture, "Female Producers," https://www .nass.usda.gov/Publications/Highlights/2019/2017Census_Female_Producers .pdf.

37. USDA 2017 Census of Agriculture, "American Indian/Alaska Native Producers," https://www.nass.usda.gov/Publications/Highlights/2019 /2017Census_AmericanIndianAlaskaNative_Producers.pdf.

38. American Farmland Trust, "Women for the Land," https://farmland.org /project/women-for-the-land/.

39. USDA 2017 Census of Agriculture, "New and Beginning Producers."

40. Sophie Ackoff et al., "Building a Future with Farmers 2022: Results and Recommendations from the National Young Farmer Survey," National Young Farmers Coalition, 2022, 71.

41. Ackoff et al., "Building a Future with Farmers 2022," 16–18.

42. Wozniacka, "Does Regenerative Agriculture Have a Race Problem?"

43. A-dae Romero-Briones, director of programs, Native Agriculture and Food Systems for First Nations, interview with the author, May 24, 2021. See also Native Land Information System, "Women's Representation in Agriculture; Greater Among Native Americans," November 2, 2020, https:// nativeland.info/blog/uncategorized/women-representation-in-agriculture/.

2. Momentum

1. Two incredible resources on this subject are *The 1619 Project: A New Origin Story*, a 2021 anthology of essays and poems created by Nikole Hannah-Jones that builds on 2019 essays published in *The New York Times Magazine*, and *Caste: The Origins of Our Discontents* by Isabel Wilkerson (New York: Random House, 2023).

2. Liz Carlisle, *Healing Grounds: Climate, Justice, and the Deep Roots of Regenerative Farming* (Washington, DC: Island Press, 2022), 81.

3. Carlisle, *Healing Grounds*, 82.

4. Emma Layman and Nicole Civita, "Decolonizing Agriculture in the United States: Centering the Knowledges of Women and People of Color to Support Relational Farming Practices," *Agriculture and Human Values* 39 (January 2022): 969. See also "The Gullah Geechee Connection," North Carolina Rice Festival, https://www.northcarolinaricefestival.org/the-gullah-geechee-rice-production.

5. William Chandler Bagley, *Soil Exhaustion and the Civil War* (Washington, DC: American Council on Public Affairs, 1942), 13.

6. Mark Bittman, *Animal, Vegetable, Junk: A History of Food, from Sustainable to Suicidal* (Boston: Mariner Books, 2022), 41.

7. Bagley, *Soil Exhaustion*, 12–17.

8. Bagley, *Soil Exhaustion*, 58, 63. I find Bagley's analysis of the connection between soil degradation, slavery, and the Civil War enlightening—but I categorically reject his racist portrayal of slaves as unintelligent and unskilled, and his too-lightly-worded objection to those who believed African slaves were biologically and intellectually suited to slavery. His quotations of slave owner descriptions of slaves are intended to prove that slavery is unproductive and unprofitable, which undermines it as an institution (a good thing). And he does describe the slaveholder as "the propagandist of the Civil War" because slaveholders, rather than admit the war was about slavery expansion, soil exhaustion, and their own profits, created narratives of hate and fear toward Blacks, false arguments about states' rights, and Southern tradition to fan the flames of war. Still, I want to make it very clear that I rely on Bagley solely for his accounts of pre–Civil War extractive agriculture and how soil exhaustion spurred the war, lessons we could learn from today.

9. Bagley, *Soil Exhaustion*, 15, 84.

10. Bagley, *Soil Exhaustion*, 87. See also Bittman, *Animal, Vegetable, Junk*, 80.

11. Facts about the post–Civil War period in this paragraph are from Henry Louis Gates Jr., "The Truth Behind '40 Acres and a Mule,'" PBS, https://www.pbs.org/wnet/african-americans-many-rivers-to-cross/history/the-truth-behind-40-acres-and-a-mule/.

12. Gates, "Truth Behind '40 Acres and a Mule.'"

13. "Conditions of Antebellum Slavery," PBS, https://www.pbs.org/wgbh/aia/part4/4p2956.html.

14. Rachel Kaufman, "In Search of George Washington Carver's True Legacy," *Smithsonian Magazine*, February 21, 2019, https://www.smithsonianmag.com/history/search-george-washington-carvers-true-legacy-180971538/.

15. Information about sharecropping in this paragraph is from "Sharecropping," PBS, https://www.pbs.org/tpt/slavery-by-another-name/themes/sharecropping/.

16. Information about George Washington Carver is from Brianna Baker, "The Land-Healing Work of George Washington Carver," Grist.org, February 12, 2021.

17. Edith Espejo, "CSA's and Regenerative Agriculture's Ties to Black History," One Earth, August 25, 2022, https://www.oneearth.org/csas-and-regenerative-agricultures-ties-to-black-history/.

18. Jess Gilbert, Spencer D. Wood, and Gwen Sharp, "Who Owns the Land?: Agricultural Land Ownership by Race/Ethnicity," *Rural America* 17, no. 4 (Winter 2002): 55.

19. Jennifer Fahy, "How Heirs' Property Fueled the 90 Percent Decline in Black-Owned Farmland," FarmAid.com, February 28, 2022, https://www.farmaid.org/blog/heirs-property-90-percent-decline-black-owned-farmland/.

20. Megan Horst and Amy Marion, "Racial, Ethnic and Gender Inequities in Farmland Ownership and Farming in the U.S.," *Agriculture and Human Values* 36 (March 1, 2019): 2.

21. Everything about heirs' property in this paragraph and the two that follow comes from Thomas W. Mitchell, "From Reconstruction to Deconstruction: Undermining Black Landownership, Political Independence, and Community Through Partition Sales of Tenancies in Common," *Northwestern University Law Review* 95, no. 2 (2001): 505–80.

22. Mitchell, "From Reconstruction to Deconstruction," 518.

23. Will Breland, "Acres of Distrust: Heirs Property, the Law's Role in Sowing Suspicion Among Americans and How Lawyers Can Help Curb Black Land Loss," *Georgetown Journal on Poverty Law and Policy* 28, no. 3 (Spring 2021): 385–86.

24. Vann R. Newkirk II, "The Great Land Robbery," *The Atlantic*, September 2019, https://www.theatlantic.com/magazine/archive/2019/09/this-land-was-our-land/594742/.

25. Horst and Marion, "Racial, Ethnic and Gender Inequities," 4.

26. Newkirk, "Great Land Robbery."

27. Ximena Bustillo, "'Rampant Issues': Black Farmers Are Still Left Out at USDA," *Politico*, July 5, 2021, https://www.politico.com/news/2021/07/05/black-farmers-left-out-usda-497876.

28. Bustillo, "'Rampant Issues.'"

29. Sophie Ackoff et al., "Building a Future with Farmers 2022: Results and Recommendations from the National Young Farmer Survey," National Young Farmers Coalition, 2022, 8.

30. Dorceta E. Taylor, "Black Farmers in the USA and Michigan: Longevity, Empowerment, and Food Sovereignty," *Journal of African American Studies* 22 (March 6, 2018): 53.

31. USDA 2017 Census of Agriculture, "Farm Producers," https://www.nass.usda.gov/Publications/Highlights/2019/2017Census_Farm_Producers.pdf.

32. Horst and Marion, "Racial, Ethnic and Gender Inequities," 5.

33. Alice Reznickova, "Lost Inheritance Black Farmers Face an Uncertain Future without Heirs' Property Reforms," Union of Concerned Scientists, June 27, 2023, https://doi.org/10.47923/2023.15127.

34. Ryanne Pilgeram, Katherine Dentzman, and Paul Lewin, "Women, Race and Place in US Agriculture," *Agriculture and Human Values* 39 (August 9, 2022): 1342.

35. Paula Dutko, Michele Ver Ploeg, and Tracey Farrigan, "Characteristics and Influential Factors of Food Deserts, ERR-140," U.S. Department of Agriculture, Economic Research Service, August 2012, 1.

36. Dutko, Ver Ploeg, and Farrigan, "Characteristics and Influential Factors of Food Deserts," 9–11.

37. Andrew Moore, "A More Diverse, Equitable Future for Natural Resources," NC State University, February 20, 2023, https://cnr.ncsu.edu/news/2023/02/a-more-diverse-equitable-future-for-natural-resources/.

3. Direction

1. Dinah Voyles Pulver et al., "'Tornado Alley' Is Expanding: Southern States See More Twisters Now Than Ever Before," *USA Today*, June 18, 2021.

2. Leah Douglas, "U.S. Corn-Based Ethanol Worse for the Climate Than Gasoline, Study Finds," Reuters, February 14, 2022.

3. Information about subsidies comes from Tara O'Neill Hayes and Katerina Kerska, "Primer: Agricultural Subsidies and Their Influence on the Composition of U.S. Food Supply and Consumption," American Action Forum, November 3, 2021, https://www.americanactionforum.org/research/primer -agriculture-subsidies-and-their-influence-on-the-composition-of-u-s-food -supply-and-consumption/#_edn4.

4. Christina Cooke, "Once on the Sidelines of Farming, Women Landowners Find Their Voices," *Civil Eats*, August 28, 2018.

5. Cooke, "Once on the Sidelines."

6. Gabrielle Roesch-McNally, "Women Non-operating Landlords; What We Are Learning About Conservation on Rented Lands," American Farmland Trust, June 1, 2020, https://farmland.org/women-non-operating-landlords -what-we-are-learning-about-conservation-on-rented-lands/.

7. Magdalena Frąc et al., "Fungal Biodiversity and Their Role in Soil Health," *Frontiers in Microbiology* 9 (April 13, 2018): 1.

8. Jane W. Gibson and Benjamin J. Gray, "The Price of Success: Population Decline and Community Transformation in Western Kansas," in *In Defense of Farmers: The Future of Agriculture in the Shadow of Corporate Power*, ed. Jane W. Gibson and Sara E. Alexander (Lincoln: University of Nebraska Press, 2019), 328.

9. Angie Carter, "'We Don't Equal Even Just One Man': Gender and Social Control in Conservation Adoption," *Society and Natural Resources* 32, no. 8 (April 3, 2019).

10. Carolyn Sachs et al., *The Rise of Women Farmers and Sustainable Agriculture* (Iowa City: University of Iowa Press, 2016).

11. Mark Bittman, *Animal, Vegetable, Junk: A History of Food, from Sustainable to Suicidal* (Boston: Mariner Books, 2022), 33.

12. Ryanne Pilgeram, Katherine Dentzman, and Paul Lewin, "Women, Race and Place in US Agriculture," *Agriculture and Human Values* 39 (August 9, 2022): 1342.

13. Megan Horst and Amy Marion, "Racial, Ethnic and Gender Inequities in Farmland Ownership and Farming in the U.S.," *Agriculture and Human Values* 36 (March 1, 2019). See also Nathan Rosenberg and Bryce Wilson Stucki, "How USDA Distorted Data to Conceal Decades of Discrimination Against Black Farmers," *The Counter*, June 26, 2019.

14. Jane W. Gibson, "Introduction: A Food System Imperiled," in *In Defense of Farmers: The Future of Agriculture in the Shadow of Corporate Power*, ed. Jane W. Gibson and Sara E. Alexander (Lincoln: University of Nebraska Press, 2019), 5.

15. USDA 2017 Census of Agriculture, "Farm Economics," https://www.nass.usda.gov/Publications/Highlights/2019/2017Census_Farm_Economics.pdf.

16. USDA National Agriculture Statistics Service, "U.S. Farm Production Expenditures, 2017," NASS Highlights, no. 2018-7, August 2018, https://www.nass.usda.gov/Publications/Highlights/2018/2017_Farm_Production_Expenditures.pdf.

17. Claire E. LaCanne and Jonathan G. Lundgren, "Regenerative Agriculture: Merging Farming and Natural Resource Conservation Profitably," *PeerJ* 6 (February 26, 2018): 1–12, e4428.

18. Stacy M. Zuber et. al., "Role of Inherent Soil Characteristics in Assessing Soil Health Across Missouri," *Agric Environ Lett.* 5 (June 29, 2020): 4.

19. Tom Philpott, *Perilous Bounty: The Looming Collapse of American Farming and How We Can Prevent It* (New York: Bloomsbury Publishing, 2020), 179.

20. Cooper County Historical Society, "Agriculture," www.coopercounty historicalsociety.org/agriculture.

21. Cody Zilverberg et. al., "Profitable prairie restoration: The EcoSun Prairie Farm experiment," Journal of Soil and Water Conservation, January 2014, 69 (1) 22A-25A.

4. Motion

1. Mark Bittman, *Animal, Vegetable, Junk: A History of Food, from Sustainable to Suicidal* (Boston: Mariner Books, 2022), 228–35.

2. USDA Economic Research Service, "Wholesaling," https://www.ers.usda.gov/topics/food-markets-prices/retailing-wholesaling/wholesaling/.

3. USDA, "Regional Food Hub Resource Guide," https://www.ams.usda.gov/sites/default/files/media/Regional%20Food%20Hub%20Resource%20Guide.pdf.

4. "Immigrant Farmworkers and America's Food Production: 5 Things to Know," FWD.us, September 14, 2022, https://www.fwd.us/news/immigrant-farmworkers-and-americas-food-production-5-things-to-know/.

5. Robin Wall Kimmerer, *Braiding Sweetgrass: Indigenous Wisdom, Scientific Knowledge, and the Teachings of Plants* (Minneapolis, MN: Milkweed Editions, 2013), 31.

6. Lawrence Mishel and Jori Kandra, "CEO Pay Has Skyrocketed 1,322% Since 1978," Economic Policy Institute, August 10, 2021, https://files.epi.org/uploads/232540.pdf.

7. Alexis Krivkovich and Marie-Claude Nadeau, "Women in the Food Industry," McKinsey and Company, November 2017, https://www.mckinsey.com/~/media/McKinsey/Featured%20Insights/Gender%20Equality/Women%20in%20the%20food%20industry/Women-in-the-food-industry-web-final-old.ashx.

8. Richard Seager et al., "Whither the 100th Meridian? The Once and Future Physical and Human Geography of America's Arid–Humid Divide. Part II: The Meridian Moves East," *Earth Interactions* 22 (2018): 1–24.

9. ChengzhengYu, Ruiqing Miao, and Madhu Khanna, "Maladaptation of U.S. Corn and Soybeans to a Changing Climate," *Scientific Reports* 11, no. 12351 (June 11, 2021): 1–10.

10. Nicola Jones, "Redrawing the Map: How the World's Climate Zones Are Shifting," *Yale Environment 360*, Yale School of the Environment, October 23, 2018, https://e360.yale.edu/features/redrawing-the-map-how-the-worlds-climate-zones-are-shifting.

11. David Wallace-Wells, *The Uninhabitable Earth: Life After Warming* (New York: Tim Duggan Books, 2020), 56.

12. Meagan E. Schipanski et al., "Realizing Resilient Food Systems," *BioScience* 66, no. 7 (July 2016): 605.

5. Amplitude

1. Diane Smith et al., "Perspectives from the Field: Adaptions in CSA Models in Response to Changing Times in the U.S.," *Sustainability* 11 (June 3, 2019): 3.

2. Daniel Prial, "Community Supported Agriculture," National Center for Appropriate Technology's ATTRA Sustainable Agriculture Program, November 2019, 13–14.

3. Jairus Rossi, "Changes in Expenditures at Local Food Market Channels in Different-Sized Communities," LFS-CFI-10, December 2021, https://lfscovid.localfoodeconomics.com/wp-content/uploads/2022/01/LFS-CFI-10.pdf.

4. Jairus Rossi, "Changes in Expenditures at Local Food Market Channels in Different-Sized Communities," LFS-CFI-2.03, May 2022, https://lfscovid.localfoodeconomics.com/wp-content/uploads/2022/05/LFS-CFI-2-03.pdf.

---

5. Andrew Carlson, Daniel Rubenstein, and Simon Levin, "New Jersey's Small, Networked Dairy Farms Are a Model for a More Resilient Food System," *The Conversation*, June 3, 2020.

6. USDA Rural Development, "Top 100 Ag Co-ops Continue Strong Performance in 2021," November 4, 2022, https://content.govdelivery.com/accounts/USDARD/bulletins/3362754.

7. Claire Kelloway and Sarah Miller, "Food and Power: Addressing Monopolization in America's Food System," Open Markets Institute, September 21, 2021, 5.

8. Carlson, Rubenstein, and Levin, "New Jersey's Small, Networked Dairy Farms."

6. Height

1. Filippa Juul et al., "Ultra-Processed Food Consumption Among US Adults from 2001 to 2018," *American Journal of Clinical Nutrition* 115, no. 1 (January 11, 2022): 211–21.

2. Two of many such studies include Bernard Srour et al., "Ultra-Processed Food Intake and Risk of Cardiovascular Disease: Prospective Cohort Study," *BMJ* 365 (May 29, 2019): 1–14, l1451, and Anaïs Rico-Campà et al., "Association Between Consumption of Ultra-Processed Foods and All Cause Mortality," *BMJ* 365 (May 29, 2019): 1–11, l1949.

3. Meghan O'Rourke, *The Invisible Kingdom: Reimagining Chronic Illness* (New York: Riverhead Books, 2022), 215.

4. "Nutritionist Tackles Serious Business of 'What to Eat,'" NPR.org, June 9, 2006, https://www.npr.org/transcripts/5474564.

5. Anahad O'Connor and Aaron Steckelberg, "Melted, Pounded, Extruded: Why Many Ultra-Processed Foods Are Unhealthy," *Washington Post*, June 27, 2023.

6. Mark Bittman, *Animal, Vegetable, Junk: A History of Food, from Sustainable to Suicidal* (Boston: Mariner Books, 2022), 242.

7. "PepsiCo Announces 2030 Goal to Scale Regenerative Farming Practices Across 7 Million Acres, Equivalent to Entire Agricultural Footprint," PepsiCo.com, April 210, 2021, https://www.pepsico.com/our-stories/press-release/pepsico-announces-2030-goal-to-scale-regenerative-farming-practices-across-7-mil04202021.

8. John Laney, "Driving Regeneration in Agriculture," Walmart.com, September 1, 2021, https://corporate.walmart.com/newsroom/2021/09/01 /driving-regeneration-in-agriculture.

9. "Regenerative Agriculture," Cargill.com, https://www.cargill.com /sustainability/regenerative-agriculture.

10. Carroll A. Reider et al., "Inadequacy of Immune Health Nutrients: Intakes in US Adults, the 2005–2016 NHANES," *Nutrients* 12, no. 1735 (June 10, 2020): 1.

11. Aysha Karim Kiani et al., "Main Nutritional Deficiencies," *Journal of Preventative Medicine and Hygiene* 63 (June 2022): E93–E94.

12. Rockefeller Foundation, "Reset the Table—Meeting the Moment to Transform the U.S. Food System," July 28, 2020, 3.

13. Donald R. Davis, "Declining Fruit and Vegetable Nutrient Composition: What Is the Evidence?," *HortScience* 44, no. 1 (February 2009): 18.

14. David Wallace-Wells, *The Uninhabitable Earth: Life After Warming* (New York: Tim Duggan Books, 2020), 63.

15. Stacey Colino, "Fruits and Vegetables Are Less Nutritious Than They Used to Be," *National Geographic*, May 3, 2022, https://www .nationalgeographic.co.uk/environment-and-conservation/2022/05/fruits -and-vegetables-are-less-nutritious-than-they-used-to-be.

16. Elena Suglia, "Vanishing Nutrients," *Scientific American*, December 10, 2018, https://blogs.scientificamerican.com/observations/vanishing-nutrients/.

17. Wallace-Wells, *Uninhabitable Earth*, 55.

18. Suglia, "Vanishing Nutrients."

19. David J. Augustine et al., "Elevated CO_2 Induces Substantial and Persistent Declines in Forage Quality Irrespective of Warming in Mixedgrass Prairie," *Ecological Applications* 28, no. 3 (April 2018): 731–32.

20. Ellen A. R. Welti et al., "Nutrient Dilution and Climate Cycles Underlie Declines in a Dominant Insect Herbivore," *Proceedings of the National Academy of Sciences* 117, no. 13 (March 31, 2020): 7273.

21. Stephan van Vliet, Frederick D. Provenza, and Scott L. Kronberg, "Health-Promoting Phytonutrients Are Higher in Grass-Fed Meat and Milk," *Frontiers in Sustainable Food Systems* 4 (February 1, 2021): 4–5.

22. Keren Papier et al., "Meat Consumption and Risk of 25 Common Conditions: Outcome-wide Analyses in 475,000 Men and Women in the UK Biobank Study," *BMC Medicine* 19, no. 53 (March 2, 2021).

23. Mauro Finicelli et al., "The Mediterranean Diet: An Update of the Clinical Trials," *Nutrients* 14, no. 2956 (July 19, 2022).

24. Ann Gibbons, "The Evolution of Diet," *National Geographic*, September 4, 2014, https://www.nationalgeographic.com/foodfeatures/evolution-of-diet/

25. Albert-László Barabási, Giulia Menichetti, and Joseph Loscalzo, "The Unmapped Chemical Complexity of Our Diet," *Nature Food* 1 (January 21, 2020): 33–37.

26. Nathan Solis, "Can't Find Sriracha? Here's Why the Shortage Is a Sign of Our Harsh Climate Reality," *Los Angeles Times*, June 26, 2023.

27. Ellie Stevens, "Georgia, the Peach State, Is Out of Peaches. Here's Why, and How Locals Are Coping," *CNN Business*, July 15, 2023.

28. See Jeff Moyer et al., "The Power of the Plate: The Case for Regenerative Organic Agriculture in Improving Human Health," Rodale Institute, 2020; David LeZaks and Mandy Ellerton, "The Regenerative Agriculture and Human Health Nexus: Insights from Field to Body," Basil's Harvest and Croatan Institute, 2021; David R. Montgomery et al., "Soil Health and Nutrient Density: Preliminary Comparison of Regenerative and Conventional Farming," *PeerJ* (January 27, 2022), e12848; and David R. Montgomery and Anne Bilke, "Soil Health and Nutrient Density: Beyond Organic vs. Conventional Farming," *Frontiers in Sustainable Food Systems* 5 (November 4, 2021), 699147.

29. Read more about Chico State's definition of regenerative agriculture in "Regenerative Agriculture 101," Center for Regenerative Agriculture and Resilient Systems, California State University, Chico, https://www.csuchico.edu/regenerativeagriculture/ra101-section/index.shtml.

30. Read more about the Regenerative Organic Alliance's definition of regenerative agriculture in "Why Regenerative Organic?," https://regenorganic.org/why-regenerative-organic/.

31. Information about the Michigan PBB contamination is from Melanie H. Jacobson et al., "Serum Polybrominated Biphenyls (PBBs) and Polychlorinated Biphenyls (PCBs) and Thyroid Function Among Michigan Adults Several Decades after the 1973–1974 PBB Contamination of Livestock Feed," *Environmental Health Perspectives* (September 26, 2017), 1–15, and "Key Points: Michigan's 1973 PBB Contamination," Community Action to Promote Healthy Environments, https://mleead.umich.edu/files/Key_Points_Michigans_1973_PBB_Contamination.pdf.

32. "Mast Cell Activation Syndrome (MCAS)," American Academy of Allergy, Asthma, and Immunology, https://www.aaaai.org/conditions-treatments/related-conditions/mcas.

33. Polyxeni Nicolopoulou-Stamati et al., "Chemical Pesticides and Human Health: The Urgent Need for a New Concept in Agriculture," *Frontiers in Public Health* 18 (July 18, 2016): 1–8.

34. Vanessa Vigar et al., "A Systematic Review of Organic Versus Conventional Food Consumption: Is There a Measurable Benefit on Human Health?," *Nutrients* 12, no. 7 (December 18, 2019): 26.

35. Ken Roseboro, "Bad Spud: GMO Potato Creator Now Fears Its Impact on Human Health," *Organic and Non-GMO Report*, October 30, 2018, https://non-gmoreport.com/articles/bad-spud-gmo-potato-creator-now-fears-its-impact-on-human-health/.

36. "Vine Removal and Desiccation," Utah Vegetable Production and Pest Management Guide, Utah State University Extension, https://extension.usu.edu/vegetableguide/potato/harvest-handling.

37. Nyasha Gumbo, Lembe Samukelo Magwaza, and Nomali Ziphorah Ngobese, "Evaluating Ecologically Acceptable Sprout Suppressants for Enhancing Dormancy and Potato Storability: A Review," *Plants* 10, no. 2307 (October 27, 2021): 2.

38. "GMO Crops, Animal Food, and Beyond," U.S. Food and Drug Administration, https://www.fda.gov/food/agricultural-biotechnology/gmo-crops-animal-food-and-beyond.

39. Kevin Loria, "Revealed: The Dangerous Chemicals in Your Food Wrappers," *The Guardian*, March 24, 2022, https://www.theguardian.com/environment/2022/mar/24/revealed-pfas-dangerous-chemicals-food-wrappers.

40. Ken Roseboro, "Grim Reaping: Many Food Crops Sprayed with Weed Killer Before Harvest," *Organic and Non-GMO Report*, February 25, 2016, https://non-gmoreport.com/articles/grim-reaper-many-food-crops-sprayed-with-weed-killer-before-harvest/.

41. Alexis Temkin and Olga Naidenko, "Glyphosate Contamination in Food Goes Far Beyond Oat Products," Environmental Working Group, February 28, 2019, https://www.ewg.org/news-insights/news/2019/02/glyphosate-contamination-food-goes-far-beyond-oat-products.

42. Luoping Zhang, "Exposure to Glyphosate-Based Herbicides and Risk for Non-Hodgkin Lymphoma: A Meta-analysis and Supporting Evidence,"

21. Giada Pierli, Federica Murmura, and Federica Palazzi, "Women and Leadership: How Do Women Leaders Contribute to Companies' Sustainable Choices?," *Frontiers in Sustainability* 3 (July 5, 2022).

8. Crest

1. Shelby Myers, "What Is the (Food and) Farm Bill and Why Does It Matter?," American Farm Bureau Federation, July 20, 2022, https://www.fb .org/market-intel/what-is-the-food-and-farm-bill-and-why-does-it-matter.

2. National Sustainable Agriculture Coalition, "What Is the Farm Bill?," https://sustainableagriculture.net/our-work/campaigns/fbcampaign/what-is -the-farm-bill/.

3. John Ikerd, "An Alternative Future for Food and Farming," in *In Defense of Farmers: The Future of Agriculture in the Shadow of Corporate Power*, ed. Jane W. Gibson and Sara E. Alexander (Lincoln: University of Nebraska Press, 2019), 389–90.

4. All information about farm consolidation in this paragraph is from James M. MacDonald, Robert A. Hoppe, and Doris Newton, "Three Decades of Consolidation in U.S. Agriculture," USDA Economic Research Service, Economic Information Bulletin 189 (March 2018): iii–ix.

5. For further reading on how industrial farmers tend to define success and the implications for rural communities, see Jane W. Gibson and Benjamin J. Gray's "The Price of Success: Population Decline and Community Transformation in Western Kansas," in *In Defense of Farmers: The Future of Agriculture in the Shadow of Corporate Power*, ed. Jane W. Gibson and Sara E. Alexander (Lincoln: University of Nebraska Press, 2019).

6. Tom Philpott, *Perilous Bounty: The Looming Collapse of American Farming and How We Can Prevent It* (New York: Bloomsbury Publishing, 2020), 80.

7. Information about subsidies comes from Tara O'Neill Hayes and Katerina Kerska, "Primer: Agricultural Subsidies and Their Influence on the Composition of U.S. Food Supply and Consumption," American Action Forum, November 3, 2021, https://www.americanactionforum.org/research/primer -agriculture-subsidies-and-their-influence-on-the-composition-of-u-s-food -supply-and-consumption/#_edn4.

8. National Sustainable Agriculture Coalition, "Inflation Reduction Act of 2022: A Deep Dive on an Historic Investment in Climate and Conservation Agriculture," August 19, 2022, https://sustainableagriculture.net/blog

22. John Hopkins Center for a Livable Future, "National Farm Bill Poll," March 2023, https://clf.jhsph.edu/sites/default/files/2023-04/national-farm-poll-results-slides.pdf.

23. California Department of Food and Agriculture, "California Agricultural Vision," https://www.cdfa.ca.gov/agvision/.

24. Julie Titone, "New Law Aims to Break Link Between Food Waste, Warming Climate," *Everett Herald*, January 15, 2023, https://www.heraldnet.com/news/new-law-aims-to-break-link-between-food-waste-warming-climate/.

25. Marissa Sheldon, "Minnesota to Increase Local Foods in School Meals," Hunter College New York City Food Policy Center, September 13, 2022, https://www.nycfoodpolicy.org/food-policy-snapshot-minnesota-usda-cooperative-agreement-local-foods/.

26. "Food Policy," City of Baltimore, https://planning.baltimorecity.gov/baltimore-food-policy-initiative/food-policy.

Conclusion

1. Barry Lopez, *Embrace Fearlessly the Burning World: Essays* (New York: Random House, 2022), 122.

2. Robin Wall Kimmerer, *Braiding Sweetgrass: Indigenous Wisdom, Scientific Knowledge, and the Teachings of Plants* (Minneapolis, MN: Milkweed Editions, 2013), 328.

Index

About the Author

Stephanie Anderson is the author of the award-winning *One Size Fits None: A Farm Girl's Search for the Promise of Regenerative Agriculture.* Her essays and short stories have appeared in outlets such as *The Rumpus, TriQuarterly, Flyway, Midwestern Gothic, The Chronicle Review,* and many others. She lives in South Florida, where she serves as assistant professor of creative nonfiction at Florida Atlantic University.